Fishing, Gone?

For Terry Robinson—
Thank you for all you do
to support angler's rights in
the U.S. and for all you do
to promote recreational angling.
Fish On!
Best,
Ed Dobbin—

THE SEVENTH GENERATION
Survival, Sustainability, Sustenance in a New Nature

A WARDLAW BOOK

FISHING, GONE?

Saving the Ocean through Sportfishing

Sid Dobrin

Foreword by M. Jimmie Killingsworth

TEXAS A&M UNIVERSITY PRESS • COLLEGE STATION, TEXAS

Library of Congress Cataloging-in-Publication Data

Names: Dobrin, Sidney I., 1967– author
Title: Fishing, gone?: saving the ocean through sportfishing / Sid Dobrin;
    foreword by M. Jimmie Killingsworth.
Description: First edition. | College Station: Texas A&M University Press,
    [2019] | Series: The seventh generation: survival, sustainability,
    sustenance in a new nature | Series: A Wardlaw book |
Identifiers: LCCN 2018045056 (print) | LCCN 2018047809 (ebook) | ISBN
    9781623497590 (ebook) | ISBN 9781623497583 | ISBN 9781623497583
    (pbk. : alk. paper)
Subjects: LCSH: Saltwater fishing—Environmental aspects—United States. |
    Saltwater fishing—Moral and ethical aspects—United States. | Fishery
    resources—Commercial vs. recreational use—United States. | Sustainable
    aquaculture—United States.
Classification: LCC SH457 (ebook) | LCC SH457 .D572 2019 (print) | DDC
    639.2/2—dc23
LC record available at https://lccn.loc.gov/2018045056

This one's for Asher and Shaia.
They are my world, my ocean.

Praise for *Fishing, Gone? Saving the Ocean through Sportfishing*

"Dobrin demonstrates his ability to not only spin a yarn but to accomplish deep research and deliver the results in digestible, entertaining fashion. Take a look at our planet's water supply through another lens. Come to understand why solutions that are so painfully obvious, remain un-instituted. . . . Knowledge is power."

—Flip Pallot, host of *Walker's Cay Chronicles*

"In his thoroughly enjoyable and enlightening style, Dobrin elucidates what it means to be a responsible, thinking angler in the twenty-first century. He systematically unravels and explains some of the thorniest issues in current saltwater fisheries management with the particular insight and sensitivity of a passionate lifelong fisherman."

—Glenn Law, editor-in-chief of *Salt Water Sportsman*

"Be forewarned: This is not 'Me and Joe Went Fishing.' . . . But I think Sid is asking the right questions, and the sportfishing community's engagement with these issues will lead to the betterment of our culture and our natural resources. Sid Dobrin is the guy on your boat to whom you'd say, 'Just shut up and fish already!' But then, later, you realize you were grateful for his company."

—Jeff Weakley, editor of *Florida Sportsman*

"Dobrin's book beautifully dissects today's major issues, from catch-shares and IFQs to the Modern Fish Act and Everglades conundrum. This exceedingly well-written text is peppered with apt quotes and historical references citing the likes of Dr. Seuss, Sylvia Earle, and Izaak Walton. If you fish saltwater, read this book for its sheer entertainment value and to become a more educated, responsible fisherman."

—Fred Garth, editor of *Guy Harvey Magazine*

" . . . an ambitious scholarly achievement . . . a manifesto by a waterman and academic that is a must read for anglers—and everyone else living on this compromised planet."

—Joe Shields, editor of *The Virginia Sportsman*

"Using his considerable gifts as a storyteller, Dobrin brings into focus all the great threats to saltwater sportfishing, including the catch-shares scandal, menhaden overfishing, red tides, dead zones, angler apathy, and government overreach. Anyone who loves the water and the creatures in it should read this book and heed its warning."

—Dave Precht, editor-in-chief of *Bassmaster Magazine*

"A must read for any saltwater angler with precise examples of why it's our responsibility to do our part in ensuring future generations can enjoy saltwater fishing. This may be the most significant book about recreational fishing written in decades."

—Tony Hart, editor of *Yak Outlaws*

"Sid has an incredible ability to match fish and fisheries issues with stories and passion for angling ethics and conserving our oceans. He has combined a lifetime of enjoyment on and in the water and shared it with us in an engaging approach that raises awareness and presents a call to action to appreciate and conserve our precious resource."

—Glenn Hughes, president of the American Sportfishing Association

"With great insight, Sid Dobrin exposes the tangled web of saltwater fisheries management and makes an impassioned case for American anglers to participate in our democratic process. Today's political climate is ripe for this type of thinking. It's time we all get on board."

—Liz Ogilvie, vice president and chief marketing officer of the American Sportfishing Association and Keep America Fishing

"*Fishing, Gone?* weaves the disparate threads of recreational and commercial fishing, angling ethics, and associated issues into an intricate web that can become an angler's guide to individual and collective empowerment. It's a fine piece of journalism and a fun read, a rare combination."

—Bryan Hendricks, outdoors editor of *Arkansas Democrat-Gazette*

"The title *Fishing, Gone?* conveys a gloomy outlook for recreational angling, but dig deeper into Sid Dobrin's new book and you'll see that it's actually a message of hope. His hope—our hope—is that the fragmented population of recreational anglers can somehow be unified in demanding responsible management of saltwater resources. *Fishing, Gone?* is a convincing call to action."

—Bruce Akin, CEO of the Bass Anglers Sportsman Society

" . . . timely, relevant, factual, and thought provoking. A must read for all who love our oceans, the fish that inhabit them, and especially for those who are passionate about saltwater recreational angling."

—Patrick Neu, executive director of the National Professional Anglers Association and president of the Future Anglers Foundation

"A thoughtful examination of the complex, ever-evolving interactions between sportfishing anglers, the commercial fishing industry, and the ecosystems we all depend on."

—David Kennedy, government affairs manager for the Boat Owners Association of the United States

"In an engaging work that mixes equal parts fish policy and fish tales, Dobrin brings to life the battle lines and trenches where the future of fishing is being debated and decided. A must read for anglers who care about their ability to go fishing, today and in the future."

—Ricky Gease, executive director of the Kenai River Sportfishing Association; Alaska Salmon Fellow

"It could be one of the most important books published this year."

—M. Jimmie Killingsworth, series editor of the Seventh Generation: Survival, Sustainability, Sustenance in a New Nature

"Who better than a university English department chair to make marine fisheries management easy to comprehend, entertaining, and thought provoking? Sid's passion for fishing allows him to clarify distinctions in commercial and recreational fisheries while showing how a new thought process is needed for both to succeed. Well done!"

—Gary Jennings, director of Keep Florida Fishing

"Blending history, politics, and storytelling, this book should inspire anglers to become more aware of our collective connection to the ocean and the risk of losing that connection if we don't become more aware of and engaged in the policy issues that affect the future of our sport."

—Mike Leonard, vice president of government affairs
of the American Sportfishing Association

"Sid Dobrin provides a wakeup call for all recreational anglers. Whether you fish with bait or artificials, if you care about the future of saltwater fishing in this country, you need to read this persuasive and highly readable book."

—John Gans, Theodore Roosevelt Conservation Partnership

"Dobrin guides readers on a spiritual and intellectual journey that leads to a profound conclusion, one that anglers must embrace, lest they forsake the very essence of angling."

—Jim Hendricks, West Coast editor of *Salt
Water Sportsman* and *Sport Fishing*

We witness a resurgence of an apocalyptic notion that the oceans will not wash away our sins but rather drown us in them.

—Stefan Helmreich, *Alien Ocean*

Contents

Foreword

If you've ever gone fishing, eaten fish, shopped for fish, read about fish, or used fish as a metaphor, this book will give you something new to think about. If you've ever been to the beach, gone surfing, taken a cruise, marveled over the mystery and expansiveness of the world ocean, watched Jacques Cousteau, or sung a sea shanty, you're also likely to find a new perspective on your experience in this densely informative yet highly readable meditation on the state of the seas.

It's the perspective of a lifelong saltwater fisherman who also happens to be a leading figure in environmental studies. Sid Dobrin chairs the English department at the University of Florida and publishes highly regarded research on English composition and rhetoric (including nature writing and eco-rhetoric). On the side, he follows the fish, his passion and second vocation. He participates in the centuries-old tradition of the Angler-Philosopher. And over the years, he has plied the trade of the outdoor writer—with an emphasis on fish and ocean. He insists on following the data as well as the fish, looking to science for guidance on the thorny problems he confronts—overfishing, overcrowding, pollution, misunderstanding of the place that fish have in our lives, and the nature of ocean ecology and its importance to life on Earth.

The knot of culture, economy, and ecology, which Dobrin acknowledges and attempts to unwind analytically while yet insisting on its existential knottiness, mirrors the problem I posed in the inaugural volume of this series. The so-called three

E's of sustainability—economy, ecology, and ethics—attract uneven emphasis, with the big E of economy too often overpowering the other E's. The problem of gauging fish catch by weight, for example, or discarding "trash fish" on piers and beaches, fails to account for the place of the fish in the overall ecology of the region and reduces them to their place in the cycle of human needs—economy, that is. The fisherman's ethic, questioned and cultivated here with passion, should not place sport and sustenance over sustainability. In fact, as Dobrin ably demonstrates, wild-catch fishing to feed the billion or more people who depend on fish as their main source of protein is simply not sustainable. The best hope, in Dobrin's view, lies in aquaculture, though fish farming as it currently stands is inadequate to supply the need and raises some serious ecological questions. The tangle of regulations and laws that attempt to govern the produce of the seas is documented and analyzed with care. Hope brims on the edges of the analysis, but the urgency of the situation has never been clearer than it is when filtered through Dobrin's context-building analysis, which is historically as well as scientifically rich.

The analysis is tempered and varied with evocative storytelling and other hallmarks of the personal essay, a form that Dobrin masterfully deploys in the service of his ideas. One element is the charming narrator in the tales about childhood adventures on the shore, boating in the deep water as an adult, fishing with mentors and old buddies. From the first page, the stories unfold through the eyes of a curious, thoughtful, but at times almost naive and childlike center of consciousness, willing to be surprised and schooled. The reader is carried along in the ongoing education of the narrator. The forays into history, popular culture, and literature are equally informative and entertaining—from the saga of the McDonald's Filet-O-Fish sandwich, to the various corporate schemes to profit from control of the ocean (brought to you by such forces as the Waltons of Walmart fame), to the rich literary tradition connected with fishing and the ocean (from Izaak Walton to Ernest Hemingway and on to Dr. Seuss).

Dobrin does not shy away from problem solving, but the book's main accomplishment is the work it does in consciousness raising. After reading it, you'll never eat a fish or book a cruise without thinking of the elaborate network of economic, ecological, and ethical questions you invoke in these seemingly simple acts. Sid Dobrin is a congenial yet demanding guide to your awakening.

—M. Jimmie Killingsworth
   Series Editor

Acknowledgments

Many people influenced this book directly and indirectly, and I am deeply grateful for the significance each has had in my thinking about fishing and in the writing of this book.

First and foremost, I wish to acknowledge and thank those with whom I fish most often and, inevitably, with whom I test the waters with ideas and theories, the ones with whom I share the camaraderie of angling and all that it encompasses:

- I offer my sincerest gratitude to Sean Morey, a.k.a. Dr. Bonefish, who, through our collaborations with Inventive Fishing and too many other projects to list here, continues to be one of the best role models I could imagine. I am a better angler and a better writer because of Sean, and for that I am endlessly indebted. I hope our adventures never end.
- I also thank the Shirtless Captain, Judd Wise of Lil' Butt Naked Fishing, whose fishing wisdom is trumped only by his generosity and friendship. Thanks, Coach, for all you share with me.
- Likewise, I can never find the appropriate words to thank Joe Marshall Hardin for all our collaborations on and off the water. When/if I grow up, I want to be Joe Hardin; he is my hero.
- Though we have not fished as often together recently, Royal Allen Brown continues to influence my thinking about fishing more than just about anyone else. I will always be grateful for his friendship and guidance.
- Similarly, I offer my sincerest gratitude to Tony Hart for his friendship, leadership, and never-ending compassion. I have

learned more about fishing from Tony in these last few years than most people know in a lifetime; for that, I am endlessly grateful.

- Though he is a dyed-in-the-wool midwesterner in every sense, I also acknowledge Aaron Beveridge's influence on me as an angler and a writer. His brilliance will never overcome his aquatic misgivings, but he is still a great guy with whom to share water time.

To each of these six, I offer my most sincere thanks for your influence on this book, for our friendship, and for all I will learn from you in the future. I raise a glass of fine bourbon to each of you with eyes toward our next times on the water.

Second, I need to acknowledge all of those who work in the recreational fishing industry and have been generous with their time in helping me think through issues taken up in this book. I offer my sincerest gratitude to all my friends and colleagues with the American Sportfishing Association: Mike Nussman, Glenn Hughes, Scott Gudes, Mike Leonard, Kellie Ralston, Mary Jane Williamson, and John Stillwagon. Likewise, I acknowledge and thank Liz Ogilvie of Keep America Fishing and Gary Jennings of Keep Florida Fishing. I also acknowledge and thank Jeff Angers of the Center for Sportfishing Policy, Patrick Neu of the National Professional Anglers Association, John Mazurkiewicz of Shimano USA, and Chris Horton of the Congressional Sportsmen's Foundation.

Finally, my sincere gratitude to M. Jimmie Killingsworth for encouraging me to write this book for the Seventh Generation series. My thanks to Stacy Eisenstark of Texas A&M University Press, who has supported this project since the outset. And particular thanks to my good friend Sam Snyder of Trout Unlimited and the Wild Salmon Center for his influential comments about the book.

Fish On.

Fishing, Gone?

1

Salvation

> The geographic magnitude of the oceans
> overwhelms our capacity to believe that
> our actions there might matter.
>
> —Dorinda G. Dallmeyer, "Incorporating Environmental
> Ethics into Ecosystem-Based Management"

We surfaced in the near-midnight dark. We had been down for about twenty minutes; the display on my wrist computer blinked twenty-nine minutes to a depth of eighty-two feet. We were low on nitrox, both of us with less than 400 psi in our tanks, but everything well within safety ranges. Everything, that is, except that there in the dark, about one hundred miles west of Tortugas Bank in the Gulf of Mexico, the boat we expected to be waiting for us on the surface was not there.

The boat, a hundred-foot steel-hulled live-aboard, had been chartered by our dive shop for a three-day spearfishing and underwater photography excursion to the Dry Tortugas, a trip we had made every year for the past few years. My dive buddy on this night dive on the second day of the outing was my friend and master dive instructor. He had been my teacher from my initial open-water certification through my professional training as a scuba instructor. A retired recon sergeant in the US Marines, Smitty had moved to Florida from Okinawa, where he had served as master dive instructor for the Marines. Smitty was one of the gunnery sergeants who reconnoitered

Grenada in the first wave of Operation Urgent Fury in the US invasion of Grenada in 1983. Smitty is the real-life guy depicted by Clint Eastwood in the 1986 movie *Heartbreak Ridge*. Smitty loves that movie. Oorah.

There on the dark surface, after a solid ten minutes of cursing the boat captain, the boat captain's mother, and the boat captain's offspring for his incompetence and failure to follow basic safety procedures, Smitty and I considered options. Any swimming would have to happen at the surface. Our near-empty tanks prohibited any sustained movement underwater, and we needed to save whatever remained in case of an emergency— that is, an emergency atop being forgotten in the Gulf of Mexico in the pitch dark. Besides, there was really nowhere for us to swim; in any direction we looked, there was just more unlit ocean. We needed to be on the surface to watch for boat lights in the dark, and we needed to remain in relatively the same area where the boat had deposited us earlier. We tethered our rigs together with about ten feet of line to avoid drifting apart. We both wore black gear, mine modeled in heroic reverence to Smitty's recon gear. In the dark, it would be easy to lose sight of one another, particularly if one of us were to doze off and stop talking. While we were at depth, I had shot a gag grouper and a nice mutton snapper, both of which dangled about six feet below me on a stringer clipped to my buoyancy compensator device (BCD). I hated to do it, but there in the dark I did not want to be tethered to that much dead bait, so I dumped the fish. I assume that Smitty did the same with whatever he had on his stringer. No need to discuss it; it just made sense to do it. We both kept our spearguns loaded with both bands.

We floated on our backs on our BCDs, masks on, snorkels in our mouths. Through my mask, I looked not at the water around me but at the sky, uncorrupted by city lights, illuminated in full darkness, a venerable cathedral ceiling. I labored to identify constellations and planets, overwhelmed by the numbers of visible stars. The view reminded me of a night sky from years before on a small island beach in the Aegean between Greece and Turkey, where I watched the stars from a blanket with a bottle of rum, a bag of fresh apricots, and a salmagundi of company. Those who have not seen the night sky from the open ocean have missed one of the most magnificent views on Earth. Those who have not seen the ocean below the surface zone have also missed magnificent things. I am a stargazer and a water gazer; I love the idea of space. Outer space. Maybe that is why I dive and fish: inner space, the depths of our planet, is as close as I can get to outer space. What was it Victor Hugo wrote? "There is one spectacle grander than the sea, that is the sky; there is one spectacle grander than the sky, that is the interior of the soul." Floating there, in the darkness all around, with the stars above, I imagined the sensation of drifting in space—which, of course, is what I was actually doing. The day they want a volunteer to be launched out there, I will be first in line. Ron Moe, in *The Starship and the Canoe*, said it best when posed the question of going into space: "I'll go," he said, "if you promise me there'll be salmon there. And cutthroat trout." I feel the same way, only I would add a few other species to the list. Me and Major Tom. For now, I will float here as debris on the surface.

"Debris," of course, is a relative term. Objects wane between being things and being debris, depending on perspective and timing. A plastic fork floating on the ocean's surface is debris, but it was a thing with human-ascribed use-value and purpose. Even before that, it was something else with ecological value: chemicals, minerals, and such. A wooden pallet floating on the water's surface is debris, having once served a function in shipping goods after having been living wood in an ecosystem, and now as debris, it reclaims its thingness—or claims a new thingness—as it serves as cover for fish and other organisms. It becomes useful again, its value renewed for the fisher—whether human, bird, or fish—who sees it not as pallet or rubbish, but as topography that might serve as a sign of game fish or food fish beneath. Its value renewed by marine organisms that make use of the cover it provides. So, I floated there in the dark on the surface of the salty sea as debris, between the boundaries of inner and outer space.

I remembered then, and again now in writing these words, being twelve years old and finding in the local public library a copy of Kenneth Brower's *The Starship and the Canoe*. I suppose if anyone had asked me before I found Brower's book what my favorite book had been, I would have been as likely to respond Thor Heyerdahl's *The Kon-Tiki Expedition: By Raft across the South Seas* as I would any other of a number of books I held (and hold) dear. The written word has influenced my life as much as has saltwater, particularly the long tradition of writings about fishing and the sea, what *Esquire* founder Arnold Gingrich so famously titled *The Fishing in Print* in his examination of five centuries of angling literature. But Brower's book gave me something new, a synthesis between a Heyerdahl-like desire for open water and a science fiction–driven fascination with interstellar travel. Brower's fusion of exploratory possibility gave me reason to contemplate my own enchantment with staring into deep blue water and deep dark space (the whole father-son focus of the book has become of more interest to me now than it was when I was twelve). I remember George Dyson's words in the book: "I'm an expert on this business of inner space and outer space, and how one is a manifestation of the other." I fantasized not only about taking George Dyson's Thoreauvian baidarka adventure through Glacier Bay and the

Inside Passage, but certainly about Freeman Dyson's theories for interstellar travel. Floating there in the Gulf of Mexico, I thought about depth, the technologies we employ to explore these two different spaces, and the vessels we need, baidarkas and spaceships, and how for that moment I was fine without either.

I am not sure exactly how many hours we bobbed around out there, but it could not have been more than about three or four before we saw the running lights on the boat long before the sun started to dissolve the darkness. Not long enough to warrant a Coast Guard search and rescue; not long enough to warrant even a made-for-TV movie. Long enough, though, for someone to realize we were not aboard and long enough for the captain to panic and return to where he left us overboard. After Smitty chased the captain into the wheelhouse—where the captain stayed with the hatch locked for the remainder of the trip for fear Smitty might fulfill one of his threats—and after Smitty and I had passed the whiskey bottle between us, after we had calmed, after we had told the story a dozen times or more, Smitty said to me, "I was never worried out there because I knew that whatever the situation was, you'd figure out how to get us out of it." To this day, I think of that as one of the best compliments anyone has ever paid me. I also confessed to Smitty that I was never worried because I knew that if anything messed with us in the water, he would kill it. I hoped he would take this as a sincere compliment. However, given that only a week prior while goofing around before a dive class I had pushed him onto a common brown water snake swimming by and he had shrieked like a small child, I knew he wasn't really as tough as he wanted people to think. Friends get to know these things.

The truth is that even though I probably should have been frightened during our small ordeal in the ocean, or at least worried, I never was. The ocean does not frighten me; it is a home to me. Floating there in the dark, with a star view better than in the Hayden, Adler, Morehead, Albert Einstein, or Pretlow Planetariums, I was relaxed, comfortable, at ease with the world. I crave that sensation of floating in the night water alone.

The ocean can be a strange place, an alien place, a wild place. Historically, we cast the ocean as the wildest nature, the untamable. Yet in the same breath, we cast the ocean as a place of salvation. Contemporary environmental conversations and oceanographic discussions describe the ocean as the place from which human salvation will likely emerge in the wake of environmental destruction; others point out that life on Earth depends on the health of the ocean: "As goes the ocean, so goes life."[1] The ocean is strange and promising all in one breath.

The ocean. Singular. The bodies of saltwater that cover the planet are connected or, more accurately, are a singular aquatic body divided only by human cartography for the sake of navigational communication, the ability to more conveniently identify location, and political claims to sovereign rights. However, such convenience invades our thought, contributing to centuries of understanding the oceans as independent bodies. Instead, we must now think not of the world's oceans, but of the world's ocean—singular—or what my friend Dan Brayton, author of the remarkable book *Shakespeare's Ocean*, points out is standard discourse in the marine sciences: "the global ocean."[2] Or what J. H. Parry, the eminent maritime historian, described in his seminal book *The Discovery of the Sea* as "the one sea": "All the seas of the world are one."[3] Unique in its oneness, the ocean is alone on this planet, and as far as we know at this moment, alone in the universe. The ocean might declare, as Rocket Raccoon does, "Ain't no thing like me, except me." It is the rarest of jewels; there is no other of its kind. As such, its value is immeasurable. Its rarity surpasses the perception of its vastness. Size is relative.

The world's ocean, though, is also complicated with the turmoil of possession. The possessive "world's" indicates the ocean as owned by the world; the world, of course, understood not to mean a global ecology, but the possession of the human inhabitants of the planet. The deep-seated cultural understanding of possession and its extension across the ocean confound our ability to think of the ocean in ways other than territorial and reveal our desire to import the land-based logic of ownership onto the fluid space of the ocean. We have records of territorial disputes over ocean access and fisheries rights dating back to at least the early 1200s. Such cultural entrenchments will be difficult to overcome. However, as I hope to show, now they must be.

It is easy, of course, to compare deep space and deep water. They are both cold, dark environments that require incredible technological apparatus for human beings to explore. Thus far, that exploration has been limited; humans have seen but a modest fraction of each. Perhaps it is our desire to overcome our terrestrial-ness that drives us to find ways to dive deeper or fly higher. Yet in our thirst to exceed our land-based lives, we have not been able to exceed our land-based thinking in fluid spaces. This consideration of the future of the ocean and of fish and fishermen begins in that space of convergence where air, water, and land meet to form surface. This book is about penetrating those surfaces and thinking differently about them, thinking differently about fish and their habitats, about what it means to fish for recreation and sport. Water and air are the media of life, and the ocean's surface is a space of media convergence and mixing. This book is about yearning to inhabit both media. This book is about what it means to live the saltwater fishing life and how the global community of saltwater anglers contributes to the future of the world's ocean. This book is not about whether environmental destruction, global warming, sea rise, or oceanic devastation is happening; those changes have been confirmed too often to ignore or doubt them. This is a book about how we now live in and with an ocean world in which drastic changes have already occurred—and continue to occur. This book is about the ocean at the end of the Anthropocene. This book is about what is next.

To live the saltwater fishing life is at once to be an angler, an oceanographer, a marine biologist, an ichthyologist, an ornithologist, a meteorologist, a geologist, an astronomer, a physicist, a chemist, a biologist, a sailor, a navigator, a wayfarer, and, mostly, a conservationist. This does not imply that anyone who fishes in saltwater automatically adopts all these facets of identity, but that those who truly live the saltwater fishing life can do so only through a dynamic engagement with the world's ocean and the complexities of the ocean's diverse ecological connections. As such, we find ourselves not as amateurs or hobbyists, but as prosumers and experts. More often now, we hear the term "citizen scientist." Perhaps, though, more than any of these, saltwater anglers are, as the patron saint of angling, Sir Izaak Walton, put it, the most contemplative of sportsmen. This, then, is a book of contemplation. This is a book about fish, fishing, and the connections saltwater fishing provides.

For those who fish, for those who are drawn to fishing and the things of fishing, the world's ocean is more than merely a location where one fishes; the ocean is not a playing field for sport. There is an inextricable bond between anglers and ocean—and not simply because that is where the fish are. To fish in saltwater is to negotiate a relationship with fluidity, to see the world differently. To fish is to touch the living world, to immerse oneself in the flow and energy of life. To fish is to crave and ache for those moments of conjunction with ocean and fish. To fish is to inhabit saltwater.

As much as anything else, my life has been defined by saltwater fishing. My addictions, desires, losses, failures, and triumphs are mixed and dissolved in saltwater, the ocean my tears of exultation and anguish. I am often asked by those who either don't fish or see fishing as a pastime for the under-cultured, "What's the biggest fish you've caught?" It's a naive question, of course, one that reduces the activity of angling not just to quantification but to comparison of unlike things. A two-hundred-pound halibut is a really big fish, but so are a seven-pound pompano, a twelve-pound bonefish, and a five-pound vermilion snapper. One cannot compare the experience of catching a halibut—no matter the size—with the experience of catching a pompano, redfish, tarpon, dolphin, or any other game fish; these are fundamentally different experiences, each with its own merits and memories. Sometimes, that is, size is overrated. For those of us who inhabit a saltwater fishing life, fishing exceeds the quantifiable. *Piscator non solum piscatur*, the old anglers' adage, claims that there is more to fishing than catching fish. For the devoted angler, this is more than truth; it is life. Sometimes, though, as I show in the pages that follow, quantification can reveal critical information that affects our lives as anglers, and as much as I resist reducing recreational saltwater fishing to numeric order, there are some numbers we can no longer ignore.

Fishing, Gone? is about ocean survival, about our imagined resilience, about the sustainability of the ocean and saltwater environments and fisheries, about the very idea of sustenance and the myth of the ocean as provider. This book is about humans surviving with the ocean and the ocean surviving with humans. This is a book about the saltwater fishing life and the role of thinking like an angler in the future of the ocean. *Fishing, Gone?* is a call to recreational saltwater anglers about what they might do to prevent a scenario in which fishing is gone. This is a book about the fear of what has not yet come to pass. *Fishing, Gone?* is about not just our activities as anglers, but the very ways we understand what it means to fish and to live the saltwater fishing life.

There are many ways in which anglers live the saltwater fishing life, just as there are many ways in which anglers know the ocean and its inhabitants. Some are personal, some are observational, some are lore and myth, some are spiritual, and some are scientific. Some are local and some are global. Some are common knowledge and some are esoteric. Some are literary and some are current events. Moreover, of course, some are flat-out wrong (my father once met a man who never fished in the ocean because he did not believe anything could actually live in something that salty). There are also things we should know about the ocean and saltwater sportfishing, if not for the sake of the saltwater fishing life, then for the sake of the very medium in which anglers invest their passions. *Fishing, Gone?* is a contemplation of possibility and of knowing. It is a purposeful move away from belief toward knowledge.

According to a study published in 2012 by the National Marine Fisheries Service (NMFS), more than eleven million people in the United States fished in saltwater in 2011.[4] Likewise, as Steven J. Cooke and Ian G. Cowx explain in their study "The Role of Recreational Fishing in Global Fish Crises" in *BioScience*, "if recreational fishing participation rates in Canada from 2000 (the most complete data available, DFO 2003) reflects global trends, roughly 11.5 percent of the world's population engages in this activity."[5] For the moment, let's stick to the United States before addressing global crises. Eleven million people. That's an impressive number. For comparison's sake, consider that following the bombing of the Boston Marathon in 2013, the membership of the National Rifle Association (NRA) surged to the largest numbers since its formation in 1871: a whopping five million.[6] The number of saltwater anglers in the United States alone is greater than the populations of 171 of the world's 247 countries. If the Saltwater Nation were an actual nation, we would be the seventy-sixth largest country in the world. Likewise, according to a 2018 report by Southwick Associates for the American Sportfishing Association, the contribution of freshwater anglers to the United States Gross National Product was $41.9 billion in 2016. The saltwater contribution was $18.3 billion with saltwater anglers spending $14 billion on saltwater angling goods and services.[7] That is an incredible amount. Comparatively, the National Football League is hoping—*hoping*—to reach annual revenue of $25 billion by 2027; it makes about $10 billion now. That is more than the entire US book and journal publishing industry makes in a year. More than the gym and fitness center industry. More than the yoga industry. More than clean-energy industries. US DVD sales take in only $23 billion. Music sales $10.4 billion. Movie sales $9.5 billion per year. Saltwater anglers contributed more than $70 billion in total economic output in 2011. Saltwater anglers support more than 450,000 jobs in the industries that orbit saltwater angling. Equally important, through federal excise taxes applied to boat fuel, license purchases, and equipment costs, as well as direct donations to conservation efforts, anglers contribute close to $1.5 billion annually toward conservation efforts and habitat restoration projects in the United States. In short, the saltwater sportfishing industry is comparatively one of the largest industries in the United States, but that economic power, that population size, and that conservation effort have not translated into a cohesive effort, agenda, or ethic. This book is an argument as to why we must now galvanize that economic, civic, and conservation power in ways we never have before.

These numbers—in particular the dollar amounts—are more than significant. In December 2016, President Obama signed into law H.R. Bill 4665, making it Public Law (PL) 114–249, the Outdoor Recreation Jobs and Economic Impact Act of 2016. The bill received strong bipartisan support in the House and Senate. It is significant because it directs the Bureau of Economic Analysis (BEA) of the Department of Commerce to assess and analyze the outdoor recreation economy of the United States and the effects attributable to it on the overall US economy. In conducting the assessment, the BEA may consider employment, sales, contributions to travel and tourism, and other appropriate components of the outdoor recreation economy. PL114–249 makes clear that we now need to account specifically for the role of outdoor recreation—such as fishing—in how we examine the entire US economy. The law thus acknowledges the national economic importance of recreational outdoor activities such as recreational fishing and the need for substantial study and data regarding their economic impacts. PL114–249 is an important acknowledgment from the federal government about the importance of outdoor activities such as recreational fishing not just in the cultural lives of Americans, but in the US economy as well.

PL 114–249 officially declares that we really can no longer understand recreational activities such as fishing as merely frivolous pastimes; they are a significant economic force. In total, the Outdoor Industry Association estimates that outdoor recreation supports 6.1 million jobs and generates over $646 billion annually. That is more jobs and revenue than the pharmaceutical and automotive industries combined.

Certainly, even before PL 114–249, outdoor industries were tracking and tabulating their economic data, but if a federal account will add to that data—whether corroborating what the industry as a whole tells us or providing new insight—having more scientifically gathered data will inevitably enhance how the industry functions and grows. This means more jobs and a more vital industry. Of course, the contribution of recreational saltwater fishing to this overall $646 billion amounts to just under 5 percent of the total; nonetheless, these are significant numbers.

I will come back to these numbers later, but for now, we must acknowledge that there are a lot of us who fish in saltwater and contribute a significant amount to the economy and to conservation efforts. We must acknowledge that recreational saltwater fishing is big business and big culture, and it is far from an esoteric activity promulgated by a handful of lowbrow rod benders. Yet there is a dramatic problem with the size of this population and its economic influence: its fragmentation. Unlike the NRA, which musters immense political power through its unified mission, saltwater anglers have no unified objective, nor any one organization through which to galvanize any significant voting or lobbying power toward conservation efforts, regulations, or other relevant issues.[8] Likewise, federal policy makers attuned to decisions regarding ocean and fisheries management have traditionally focused on commercial harvesting issues. Granted, in 2013 the *State of the Coast Report* by the National Oceanic and Atmospheric Administration (NOAA) identified that in 2011 the commercial fishing industry supported nearly one million jobs in the United States and generated $32 billion in income. That is about twice as many jobs as were supported by recreational saltwater fishing, and $5 billion more in generated income. Likewise, the commercial fishing industry is much better organized in using its lobbying and voting power to keep federal eyes turned toward its needs and developing a rhetoric of need around its industry.

We have to keep in mind, too, that it is not just political fragmentation or objective disagreement that prevents galvanization among saltwater anglers; it is also diversity in fishing approaches, targeted species, local environments, and economic access. Hand lining for mangrove snapper from a bridge in South Florida is very different from trolling the Gulf Stream for marlin. Surf casting for bluefish on North Carolina's Outer Banks is remarkably different from rock fishing in Southern California. Kayak fishing for redfish along the Texas Gulf Coast is different from kayak fishing for sailfish in South Florida. Flats fishing for tarpon and bonefish in the Florida Keys is different from flounder fishing off the New Jersey coast. Halibut fishing in Alaska is different from pier fishing for croaker in South Carolina. Trolling for wahoo in Hawaii is different from fishing for striper in Rhode Island. And so on. Every kind of saltwater fishing requires different equipment, different economic investment,[9] different personal investment, different abilities to interact with the environment, and above all else, a different way of thinking about fishing.

Anglers are inherently philosophical, developing ethical and bioethical approaches to the animals and environments with which they interact—admittedly, often tacitly. The decisions anglers make require careful consideration in fluid flux. Decisions, for example, to practice catch-and-release angling versus catch-and-kill angling require not just simple decision making about the use-value of a fish in a given moment, but a more complex value assessment of life in general. The meat angler, whose objective is to fill coolers with as much fish meat as possible, approaches the world from a different philosophy than does the angler who will harvest only one fish for the pan. The catch-and-release angler who fishes with bait thinks differently about fishing than the fly-only catch-and-release angler. Such differences only scratch the surface of philosophical differences among recreational anglers. I note these evident differences not to ascribe a value system or hierarchy within recreational angling, as some might want, but to begin to identify the broad discrepancies of principles that one must navigate within the sportfishing community. It would be hypocritical for me to deem any one belief about recreational fishing superior to another, not only because my own practices vary from context to context, but because nearly all anglers will at some time in their angling life face an ethical dilemma about how to establish and adhere to their own angling values. As a matter of disclosure, I fish with bait, artificials, and flies. I practice catch-and-release more often than not, but I also practice catch-and-eat when the situation is right. Most of my decisions are steered by the immediate context. Which species am I pursuing? Where am I geographically? What is the state of the fishery in this place? What do local regulations dictate? What does local knowledge reveal? Is the hooked fish likely to recover from the release? And so on. To address any one of these and other questions ethically, however, anglers must be alert to the constantly changing situation in which they fish. That is to say, there are never easy answers to ethical questions.

A handful of years ago, I killed a bluespotted grouper (*Cephalopholis argus*)—a roi—in Hawaii. The iridescent blue of the fish was mesmerizing, even as it lay dying on the ice in my cooler. My intentions had been culinary in origin (and thus, I told myself, honorable), as my Floridian heritage taught me to understand "grouper" as synonymous with "grilled," "blackened," or "fried." However, a friendly kama'aina who asked to see my catch warned me against eating the fish, repeating that it was *kapu* and would make me sick if I ate it.

Hawaiian roi, it turns out, often carry the ciguatera toxin, which can cause serious gastrointestinal and neurological illness—and sometimes death—in humans. I left the roi in a gas station garbage can. The sight of the fish atop a rubbish bin filled with food wrappers and plastic drink bottles left me dispirited, unsure of my choices, and agitated by my own ignorance.

I held on to that self-directed disappointment for quite a while, often thinking of that grouper in the garbage can, its prismatic ultramarine faded in death, the fish rendered nothing more than trash by my hand. I sought penance in learning more about the roi. What I found, though, was not absolution, but rationalization. I learned that the fish is not native to Hawaii. According to a 1987 article by John E. Randall, published in the *Bulletin of Marine Science*, in 1956 the *Hugh M. Smith*—a 128-foot, ex-Navy auxiliary outfitted in 1949 to conduct oceanographic studies and semicommercial-scale tuna fishing— transported 571 small roi from Moorea in the Society Islands to the Hawaiian Islands. According to Randall, "one lot of 171 was released off Browns Camp, O'ahu, and the remaining 400 off Keahole Point on Hawaii." By 1958, evidence that the roi had begun spawning emerged. Around this same time, researchers introduced twenty-one species (including four other grouper species and four snapper species) to Hawaiian waters to help develop local, harvestable fisheries. Of these, seven established breeding populations, but only the roi and two other species adapted to the Hawaiian reefs and still maintain populations in the area.[10]

The roi, taking advantage of the welcoming environment, have grown to be the preeminent predator of Hawaii's reefs and have expanded their population size beyond anticipated numbers, changing the reef's ecosystem dramatically. Hawaii's Division of Aquatic Resources (DAR) and the Nature Conservancy report that the roi population of Hawaii's reef system has increased more than fifteen-fold in the last thirty-five years. Research circulated by the DAR and conducted at the University of Hawaii has shown that in the very region from which I took the garbage-can grouper, this massive population of top predators devours more than ninety tons of reef fish annually, about 8.2 million fish.[11] The roi is now considered an invasive species, and local spearfishermen hold "roi roundups" that target the roi in an attempt to reduce the size of the population.[12] In a word, then—"invasive"—I was granted rational absolution for the killing of the grouper, as that one word and the history of the fish shielded me from my own initial ignorance, conceding not forgiveness, but situational necessity. I had (admittedly unintentionally) assisted in the local effort to conserve a native ecosystem, the life value of the fish altered as its death evolved from unnecessarily killed to killed for the greater good.

Reductively, we could say that recreational fishing boils down to hook, line, ocean, fish, and angler, but such an over-simplification is tantamount to saying that religion is about a person and a god. It is just not that simple. The diversity of fishing and anglers, like the diversity of religions and believers, is much more complex (and, let's face it, sometimes indistin-guishable, as many of us place a tremendous amount of faith in fishing) and much more deeply influential as to how each person sees the world and lives the saltwater fishing life. That is, our relationships with the ocean, with fish and fishing, are always more complicated than simply saying "I fish."

Despite the diversity, though, given the size of the salt-water angler population, one might think the potential energy of such a mass of individuals with a common interest would be better suited for collaboration and cooperation. In 2013, under the leadership initiatives of the American Sportfishing Association (ASA), sportfishing industry leaders formed the

Commission on Saltwater Recreational Fisheries Management, which delivered its first report in the spring of 2014. Cochaired by Johnny Morris, founder and CEO of Bass Pro Shops, and Scott Deal, president and cofounder of Maverick Boats, the Morris-Deal Commission, through a report titled *A Vision for Managing America's Saltwater Recreational Fisheries*, called for a national policy on recreational fishing. On April 2, 2014, NOAA announced that it would begin to develop a national policy on recreational fishing, though it was initially unclear whether this was a response to the commission's report. In December 2014, the National Marine Fisheries Service, a division of NOAA, released the first draft of the policy statement for public comment. The draft statement reflected closely the language of the commission report and aligned its policies with the Magnuson-Stevens Fishery Conservation and Management Act (MSA). Following a period for public input on the draft, on February 12, 2015, at the Miami International Boat Show, NOAA Fisheries administrator Eileen Sobeck announced the new National Saltwater Recreational Fisheries Policy (NSRFP), a landmark moment in recreational saltwater fishing.

I must have read the policy a dozen times in the first few days after its release, mollified by its concision and directness. I wanted to feel its history in the way I imagined the gathering at the American Museum of Natural History felt on June 7, 1939, when the International Game Fish Association was officially launched, providing for the first time an internationally recognized code of ethics for saltwater anglers. I was pleasantly surprised to see that the federal policy acknowledged the importance of the Morris-Deal Commission report, and encouraged to see the introduction to the policy recognize the economic and cultural differences between recreational fishing and commercial fishing, and the growing influence of

recreational anglers. Yet at the same time, this brief acknowledgment (which had not appeared in the previously circulated
first draft) brought to mind a question of fundamental
importance. In 1994 voters in Florida—the self-proclaimed
capital of saltwater sportfishing—passed a constitutional
amendment to ban inshore gill netting. The ban, which was
popularly known as the "Save Our Sealife" (SOS) campaign, is
the only initiative in Florida history to have found its way onto
the ballot without the use of paid petition gatherers; more than
520,000 voters signed the required petitions to make it a ballot
issue—only 429,428 signatures were needed. In fact, support
for the amendment manifested in 72 percent of Florida's voters
calling for its approval—the greatest victory for any proposed
amendment in Florida's history. Part of what made the SOS
campaign such an important event was that it was initiated
and led by Karl Wickstrom, publisher and editor of *Florida
Sportsman* magazine, one of the best-selling fishing magazines
in the country. Wickstrom began his campaign with a simple,
informative column about the ecological damage of inshore
gill netting. In the years following the passing of the landmark
amendment, Florida anglers would be quick to point out that
the amendment revitalized inshore fishing. But perhaps more
important, Wickstrom's readers and anglers across the state
would also begin to remember the ban as a moment when
anglers voiced their frustration with the government's inability
to manage and protect marine fisheries.[13] That is to say, the
1994 Florida net ban should be understood as a powerful
moment when citizen anglers joined under a common banner
not only for marine conservation, but as a public proclamation
of voter power and voter dissatisfaction with governmental
approaches to protecting saltwater fishing environments and
populations. As I will show later, the SOS net ban has had

more than ecological impact; those who supported the ban and those who opposed it regularly cite it as the catalyst for contemporary policies and practices. It has also directly influenced the ways in which regulatory battles are fought over fisheries management.

Twenty-one years after the SOS amendment, I am grateful for the federal government's consent to finally develop policy for recreational saltwater anglers and grateful to see the willingness to acknowledge the efforts of the Morris-Deal Commission. I wonder, too, about the differences between the 1994 Florida action and the new NMFS policy, particularly because of what has unfolded since the initial introduction of the national policy. A longtime South Florida investigative reporter initiated the SOS movement, and citizens without big money backing spearheaded the effort. Voters understood the proposed amendment as a statement of dissatisfaction with government fisheries management and conservation efforts. On the other hand, high-dollar industry inspired the NMFS policy, and big government promulgates its circulation. Big money plus big government is generally not a formula for which the conservation-minded citizen hopes. As I said, though, I am encouraged by the release of the new policy and the acknowledgment of the Morris-Deal Commission's influence. You see, mustering nearly three million voters in Florida twenty-one years ago was certainly a critical moment in the history of saltwater sportfishing, but now the world's ocean requires more. It requires that in the United States we mobilize as many of the eleven million saltwater anglers as possible (and thirty-eight million freshwater anglers, as well), and that we begin to think not only about local fisheries issues but specifically about the ecological intertwining of all fisheries issues as part of a larger, global oceanic conservation. I see the role of the Morris-Deal Commission, the NMFS Saltwater Recreational Fisheries Policy, the SOS amendment—and, as I will address, the 2017 addition of the Modernizing Recreational Fisheries Management Act of 2017—as inseparable events in a larger project not just to protect the world's ocean and fisheries, but to reenvision the ocean and fisheries in a new way. All of these are nodes in a more complex network that we must necessarily learn to navigate in order to think forward regarding ocean conservation. Such work and such understanding must be grounded in scientific knowledge, in the kind of citizen action we witnessed in Florida, in the innovation of the industry, and in the collaboration of government, science, citizen, and industry.

My vision, however, may be idealistic, hopeful, and—perhaps—unrealistic. In the months following the release of the national policy, several events unfolded that revealed not a unified saltwater angler nation but a knotted, conflicted population caught in a complex tussle of fisheries management, federal policies, state efforts, conservationist agendas, lobbyist politics, and industry pressures.

Sobeck's announcement at the boat show was a strategic move to emphasize NOAA's support of the boating and recreational fishing industries—a relationship that must continue to strengthen. This relationship had historically taken not the back seat to NOAA's relationship with the commercial fishing industry, but an entirely different vehicle. The policy outlined six guiding principles:

- Support ecosystem conservation and enhancement.
- Promote public access to quality recreational fishing opportunities.
- Coordinate with state and federal management entities.
- Advance innovative solutions to evolving science, management, and environmental challenges.
- Provide scientifically sound and trusted social, cultural, economic, and ecological information.

- Communicate and engage with the recreational fishing public.

Two months following the release of the policy, NOAA released a follow-up document identifying an implementation plan for the new policy, organized by way of these six principles. However, at the same moment that the policy implementation was being distributed, longtime concerns about red snapper fisheries management in the Gulf of Mexico were coming to a head, as erratic federal policies left recreational anglers and state management agencies frustrated with discrepancies between red snapper population data and access to red snapper harvest.

According to NOAA research, red snapper populations have improved in the Gulf of Mexico.[14] Yet NOAA and NMFS continued to reduce the amount of time recreational anglers could catch red snapper in federal waters, highlighted by a ridiculously short and historically low season of just three days with a two-fish limit in 2017 before the Department of Commerce intervened and brokered an exchange of an extended state water season. Recreational anglers had already complained when NMFS imposed a nine-day season in 2013 (which included only one weekend, preventing many recreational anglers who could get out only during the weekends from participating in the limited harvest) and a ten-day season in 2015. A decade prior, the season had lasted six months with a four-fish bag limit.

Driven by frustration over access and the lack of consistency, the directors of the fish and wildlife agencies of the five Gulf States—Alabama, Florida, Louisiana, Mississippi, and Texas—issued a joint statement on March 13, 2015, announcing that they had "developed a framework for cooperative state-based management of Gulf red snapper."[15] The five states proposed to work together to develop a new, independent management body called the Gulf States Red Snapper Management Authority (GSRSMA), which would comprise the principal fisheries managers from each Gulf State. Most important, red snapper management would no longer reside with the Gulf of Mexico Fishery Management Council (GMFMC), one of eight regional fishery management councils established by the Magnuson-Stevens Fishery Conservation and Management Act of 1976. The GMFMC develops fisheries management plans in the Gulf within a two-hundred-mile limit from where state waters end—a region also known as the Exclusive Economic Zone (EEZ). This region is overseen by the United Nations Convention on the Law of the Sea, which confers sovereign rights to countries over adjacent waters. Within a few days of the five-state announcement, representatives of the commercial fishing industry began to respond by resisting the very idea of state-based management. Of course, we anticipated commercial fishing's rebuff of the idea. After all, in 2006 the GMFMC had developed the Red Snapper Catch Share Program, which allocated 51 percent of the total red snapper harvest to commercial industry. According to the Coastal Conservation Association (CCA), fewer than four hundred commercial operators now own 51 percent of the entire red snapper fishery, and there are plans to allocate about another 20 percent of the catch to charter and for-hire business, leaving only a 29 percent share to recreational anglers. In chapter 5, I look more closely at catch-share programs, the privatization of the ocean, and their effects on recreational anglers.

On March 20, 2015, the presidents of the American Sport-fishing Association, the Billfish Foundation, the International Game Fish Association, the Congressional Sportsmen's Foundation, the Coastal Conservation Association, the Guy Harvey Ocean Foundation, the Center for Coastal Conservation, and the National Marine Manufacturers Association signed a letter to the governors of the five Gulf States supporting the state-based Gulf red snapper management model, citing the strength of the sportfishing community as economic contributors to each state and as significant contributors to conservation efforts in each state. Each of these eight organizations represents the sportfishing industry, ocean conservation efforts, and political lobby groups. Working together to support the proposed GSRSMA effort, this coalition represents one of the first times that traditionally disparate organizations have worked together in support of both state-based policy and, more directly, in favor of recreational angler rights.

On this same day, Alaska congressman Don Young introduced a bill to reauthorize the Magnuson-Stevens Fishery Conservation and Management Act (MSA)—originally known as the Fishery Conservation and Management Act of 1976. Earlier versions of the law had established the eight regional fishery management councils, including the GMFMC. Young's revision to the law was designed to provide increased flexibility for the councils in developing management policies. The original act specified the need to base policy about harvest limits on scientific evidence in order to end overfishing and rebuild fish stocks. The new revisions permit the councils to consider local circumstances, such as a community's economic situation, when establishing harvest limits. The added flexibility also encourages the gathering of further scientific data in regions where data are lacking.

I should note, too, that a significant part of the problem with the councils' management of fisheries policies from the perspective of recreational anglers has been the composition of each council. NOAA appoints council members; they are not elected, nor are there representation distribution requirements. Historically, commercial interests dominated—if not monopolized—the councils' makeup. Only recently have recreational interests begun to find seats on the councils, and still only in very limited numbers. As Mike Leonard, vice president of government affairs of the ASA, explained to me in a conversation about the councils:

> Regional Fishery Management Councils have historically been dominated by commercial fishing interests. While the overall make-up has been shifting slightly in recent years, there are still far more commercial fishing representatives than recreational fishing representatives on the Councils. Nowhere is this more evident than in the southeastern part of the country. Saltwater recreational fishing in the South Atlantic and Gulf of Mexico is huge for the recreational fishing industry. Over half of all saltwater fishing trips in the U.S. take place in the southeast. According to NOAA's most recent economics report, there are almost twice as many jobs supported in this region by recreational fishing (165,118) than commercial fishing (93,916). Despite all the jobs supported by recreational fishing, out of the 34 seats combined on the South Atlantic and Gulf of Mexico Fishery Management Councils, only four are held by what could be considered private recreational fishing representatives. It is very difficult for recreational fishing to get a fair shake when our community is barely allowed at the table.[16]

Thus, any attempt to shift fisheries management from the councils' authority to other oversight—such as a state-based system, as the coalition proposed, or Young's revision to Magnuson-Stevens, which would provide greater flexibility in council authority—would inevitably be seen as an assault on commercial fishing's hold over the councils.

On April 1, 2015, Chris Brown, president of Seafood Harvesters of America, issued a statement criticizing the five-state plan, arguing that "the scheme will open the door to devastating overfishing and set a dangerous precedent that could unravel the responsible management of America's fisheries from under the landmark Magnuson-Stevens Act." Brown goes on to argue that commercial fishermen have strictly adhered to their quotas while recreational fishermen have violated theirs in six of the last eight years. Brown writes: "A plan designed to take quota from the commercial sector and increase the recreational quota even further will likely open the door to more unsustainable overfishing. Recreational fishermen should focus on creating an accountable fishery for themselves, rather than using politics to reallocate access to American Red Snapper." According to Brown, the "purpose of the proposal is to increase the proportion of American Red Snapper that will be allocated to recreational fishermen who have far less scrutiny and accountability of their catches than commercial fishermen." He concludes: "Commercial fishermen from the Gulf of Alaska, to the Gulf of Maine, and south to the Gulf of Mexico have expressed deep concerns that abandoning the council system could set a dangerous precedent for unraveling fishery protections around the country—costing them a way of life and endangering the future of wild caught, sustainably harvested American seafood."[17]

Of course, Brown's arguments are thinly veiled ways of saying that making any changes to the management system would result in commercial stakeholders losing control over the councils and thus their influence in directing management policies that benefit the commercial sector. The concepts of "access allocation" and "way of life" are central to the commercial fishing rhetorical approach for maintaining control over large portions of harvest access. As concepts, they are deeply problematic; as philosophies on which to base arguments about fisheries management rights, they exemplify core problems with how we tend to think about fish populations, fishing activities, and "rights" to take fish from the ocean. I will return to these ideas in greater detail later; for now, the significant factor is that portions of the commercial fishing industry assume that control of access to American red snapper by way of federal policy is indicative of their view of all allocated fisheries management and of their response to an emerging recreational fishing voice in policy development. Their continued support of maintaining the MSA without revision is an attempt to maintain commercial control of fisheries populations as the status quo.

On April 30, 2015, the Congressional Sportsmen's Foundation hosted a breakfast briefing on Capitol Hill to address "The Solution to Gulf of Mexico Red Snapper: State-Based Management." The breakfast, sponsored by many of the same groups that signed the letter to the Gulf State governors, focused on the inability of the NMFS and the Gulf of Mexico Fishery Management Council to effectively manage the red snapper fishery and the resulting need to move management authority to the Gulf State fish and wildlife agencies. According to reports from the breakfast meeting, "the current use of a 'one size fits all' commercial fisheries management model for Gulf red snapper has resulted in the shortest recreational fishing season in history last year, despite a healthy and rapidly growing population of Gulf red snapper."[18]

The same day as the breakfast, the House Committee on Natural Resources approved Young's bill (H.R. 1335) in a divided roll call vote of 21–14, acknowledging the priorities of the recreational fishing community as articulated in the Morris-Deal Commission's report. The bill, as it was sent forward, was designed to "provide flexibility for fishery managers and stability for fishermen, and other purposes."[19] The day following the bill's passing, the ASA issued a statement applauding the move of the bill to the full House as a sign of the influence the saltwater recreational fishing community was finally having in fisheries management laws and the ability to inform changes to Magnuson-Stevens. The ASA identified the influence of the Morris-Deal Commission report on the proposed MSA revisions and thanked Congressman Young for proposing changes to the nation's marine fisheries laws to account for recreational saltwater fishing interests. Citing support from Whit Fosburgh, president and CEO of the Theodore Roosevelt Conservation Partnership; Jim Donofrio, executive director of the Recreational Fishing Alliance; and Jeff Crane, president of the Congressional Sportsmen's Foundation, the ASA continued to establish a vision of saltwater anglers as beginning to forge a coalition approach to common management issues. The statement likewise commended Congressman Jeff Duncan of South Carolina for submitting an amendment to "prompt a review of quota allocations in fisheries in the South Atlantic and Gulf of Mexico with both a commercial and recreational component."[20] The ASA noted, too, that Congressman Garret Graves of Louisiana's proposed amendment to transfer the management of Gulf of Mexico red snapper to the five Gulf States had been excluded, despite widespread agreement that Gulf red snapper management needed significant change. I have been fortunate to speak with Representative Graves a few times while writing this book, and I am grateful for his insight and leadership in addressing not just Gulf red snapper and catch-share issues, but also recreational angling rights in general.

To confuse matters, though, when the House Committee

on Natural Resources sent H.R. 1335 forward to the full
House for a vote, the Office of Management and Budget in the
Executive Office of the President of the United States issued a
Statement of Administration Policy responding to the proposed
MSA revisions in H.R. 1335. The administrative statement
includes direct criticism of the state-based red snapper manage-
ment plan, indicating that H.R. 1335

> would also severely undermine the authority of the Gulf of
> Mexico Regional Fishery Management Council by extend-
> ing State jurisdiction over the recreational red snapper
> fishery to nine miles in the Gulf of Mexico. This pro-
> posed extension of jurisdiction would create an untenable
> situation where recreational and commercial fishermen
> fishing side-by-side would be subject to different regulatory
> regimes. Absent an agreement among the States as to how
> to allocate recreationally-caught red snapper, the bill would
> encourage interstate conflict and jeopardize the sustainability
> of this Gulf-wide resource.[21]

The statement goes on to urge Congress to support the previous
versions of the MSA and changes proposed by NOAA rather than
the H.R. 1335 revisions. The statement concludes adamantly: "If
the President were presented with H.R. 1335, his senior advisors
would recommend that he veto the bill." Clearly, the administration was less concerned with the flexibility and local management
opportunities provided in H.R. 1335 than with maintaining
centralized management. Ultimately, this distinction between
local management versus standardized national management
must be considered in depth if we are to move toward sustainable
fisheries management that accounts for both recreational harvest
and smaller, independent commercial harvest.

On May 21, House Resolution 274, "providing for consideration of the bill (H.R. 1335) to amend the Magnuson-Stevens Fishery Conservation and Management Act to provide flexibility for fishery managers and stability for fishermen, and for other purposes,"[22] was approved by a House of Representatives vote of 237–174 (21 representatives did not vote), moving the revisions one step closer to approval. Interestingly, the vote was divided along party lines, with Republicans providing 236 yeas and 1 nay, and Democrats 1 yea and 173 nays. Eight Republicans and 14 Democrats did not vote.

The next day—three days after the White House administration made its statement—U.S. Ocean Conservation, a Pew Charitable Trust, began denouncing the revisions to the MSA as ignoring the achievements of the act in prior years, claiming that the revisions moved fisheries management in the wrong direction. An official statement from Pew decried, "Congress should reject this legislation and instead support provisions that protect the nation's fish populations and the millions of Americans who depend on them."[23] Part of Pew's argument against the revised version was directly aligned with the administrative statement's criticism of the proposal to extend Gulf State jurisdictions for red snapper management from three to nine miles for recreational fishing and not commercial fishing. Pew's statement focused on forage fish conservation, bycatch minimization, habitat protection, and the need to use scientific data to inform policy making. Pew's critique also left many people conflicted in trying to understand how trusted science-based research organizations could be so adamantly aligned with industrial commercial harvest management policies.

On June 1, 2015, under the provisions of H.R. 274, H.R. 1335 was presented to the US House of Representatives and, after minimal debate, was passed by a recorded vote of 225–152. The breakdown, again following party lines: 220 Republican yeas, 3 Republican nays, 5 Democratic yeas, and 149 Democratic nays.

On June 2, 2015, H.R. 1335 was received in the Senate, read twice, and referred to the Committee on Commerce, Science, and Transportation, which was composed of 25 senators and led by committee chairman John Thune (R-SD) and ranking member Bill Nelson (D-FL). The committee was described as being "composed of six subcommittees, which together oversee for the vast range of issues under its jurisdiction. These issues range from communications, highways, aviation, rail, shipping, transportation security, merchant marine, the Coast Guard, oceans, fisheries, weather, disasters, science, space, interstate commerce, tourism, consumer issues, economic development, technology, competitiveness, product safety, and insurance."[24] On January 3, 2017, the bill was introduced to the Senate as H.R. 200: Strengthening Fishing Communities and Increasing Flexibility in Fisheries Management Act.

At the same time the red snapper debacle was unfolding, two other events were materializing that would affect the federal picture of recreational fishing. First, the National Park Service (NPS) issued a new general management plan for Biscayne National Park, closing more than ten thousand acres of Biscayne Bay to recreational fishing but not to other water activities such as diving and snorkeling. Many viewed this move as dubious, as the NPS refused to consult with the Florida Fish and Wildlife Conservation Commission (FWC) before announcing the closures. Likewise, the move clearly favored access to public waters by one recreational industry over another. Nor did the NPS provide any scientific evidence that such closures were proven to restore reefs. In response to the closure, Florida representative

Ileana Ros-Lehtinen proposed a bill that would require the NPS
and the Office of National Marine Sanctuaries to obtain permission from local state wildlife agencies before implementing a
fishing closure. The bill H.R. 3310, Preserving Public Access to
Public Waters Act, was introduced to Congress in July 2015 and
was simple in its proposal:

> The Secretary of the Interior and Secretary of Commerce
> shall not restrict recreational or commercial fishing access to
> any state or territorial marine water or Great Lakes waters
> within the jurisdiction of the National Park Service or the
> Office of National Marine Sanctuaries, respectively, unless
> those restrictions are developed in coordination with, and
> approved by, the fish and wildlife management agency of
> the state or territory that has fisheries management authority
> over those waters.[25]

On August 31, 2015, the bill was referred to the Subcommittee
on Water, Power and Oceans. There has been no action on this
bill since, and the closed areas of Biscayne Bay remain off-
limits to anglers.

Second, the federal government began to explore the possibility of establishing two marine national monuments in New
England waters. Proposed by a coalition of some of the country's largest environmental groups, including the Conservation
Law Foundation, Natural Resources Defense Council, Pew
Charitable Trusts, Earthjustice, Environment America, and
other groups, the marine national monuments would protect
two areas: Cashes Ledge in the Gulf of Maine and the New
England Coral Canyons and Seamounts.

National monuments differ from national sanctuaries in
that the sanctuaries can be designated either by congressional

action or by the NOAA, but monuments can be designated only by presidential proclamation, as indicated by the Antiquities Act of 1906. Traditionally, presidents used this authority to designate public lands as national monuments, but recent presidents have also used the act to establish marine national monuments. Prior to the proposal of the two new marine monuments, there were thirteen national marine sanctuaries, but only four marine national monuments—all in the Pacific: the Papahānaumokuākea, Midway Islands, Marianas Trench, and Rose Atoll.

Supporters of the proposed New England monuments, which would be located east of Cape Cod, backed a plan for the monuments to be designated as "fully protected" regions, meaning that they would be closed to recreational fishing, as well as commercial and industrial harvesting. Fortunately, in 2014, when President Obama expanded the Pacific Remote Islands Marine National Monument through Executive Order 12962, he amended the details of the EO by adding EO 13474, allowing for recreational fishing and acknowledging that recreational fishing is a sustainable use of a public resource. The president's move to acknowledge recreational fishing as a sustainable activity confused the message his advisers had sent regarding the MSA revisions and their public recommendation that the president veto H.R. 1335. The presidential advisers claimed that recreational fishing was not a sustainable activity, counter to the president's own move in the amended Executive Order.

On Thursday, September 15, 2016, President Obama designated the Northeast Canyons and Seamounts Marine National Monument as the first national marine monument in the Atlantic. He made the announcement at the Our Ocean Conference held in Washington, DC, and hosted by

Massachusetts senator John Kerry. As he had done in his 2014 expansion of the Pacific Remote Islands Marine National Monument, President Obama again included recreational fishing as a sustainable activity permitted within the boundaries of the monument.

In July 2015, the policy-related events of the previous months propelled an undercurrent throughout the International Convention of Allied Sportfishing Trades (ICAST), a remarkably large trade show for the sportfishing industry run by the ASA. Amid the seminars and conversations addressing the Biscayne closures, the red snapper debacle, the MSA revisions, and other conservation-related issues, I spoke with representatives of the ASA, Keep America Fishing, NOAA, US Fish and Wildlife Service, and CCA as well as magazine editors, industry leaders, and leaders of conservation organizations. I had recently begun voicing my thoughts about fisheries policy through an online video series called *From the Head*,[26] which is published on the Inventive Fishing web pages by way of YouTube. The videos had gained a small following among groups invested in the sportfishing industry, and therefore many of those involved in such industry-level discussions were willing to talk with me. As energizing as I found those three days of conversations, I began to worry that my conversations and those held throughout the industry were insular. That is, if there are eleven million saltwater anglers in the United States, then why were so few talking about Magnuson-Stevens, about closures, about ethanol, about red snapper, or about any one of the dozens of pressing issues tied directly to sportfishing and conservation? It became clear to me that as well-meaning and important as the efforts of the many conservation-minded organizations within saltwater sportfishing were, most of those efforts did not reach the average recreational angler. At the time, the ASA and others were encouraging anglers to contact their congressional representatives to voice their support for ratification of the MSA revisions, but it became clear that at most a few thousand anglers were picking up the phone, and most of those were tied directly to the industry. How does one speak to and with a population eleven million strong with an independent voice guided neither by industry objectives nor regulating agency agendas? This is, in part, one of this book's catalyst questions.[27]

For quite a while now, not a day has passed in which news about conflicting approaches to fisheries management has not made its way into my inbox. Questions about fisheries policies and access to fish seem to unfold at a rapid pace. It is within this political and cultural climate that *Fishing, Gone?* intervenes. In the midst of the arguments and proposals, I find myself thinking about the significant disconnect between the policy and the fish, the angler and the agencies, the discussion and the experience. And it is there that my thoughts dwell, in the experience of fish and ocean and human, in the practice of fishing and in the very idea of fishing. From the time I was about ten years old through my college years, I kept a notebook in which I scrawled memorable quotes about fishing and the sea that I discovered in my reading. I have always been an avid reader and have always eagerly read all I can about fishing and the ocean. My reading of policy is undeniably influenced as much by the philosophies I have extracted from books, magazines, and journals over the years as by my years of on-the-water experience. Though not now considered one of his better-known novels, George Orwell's *Coming Up for Air* was critically acclaimed when first published in 1939, just before the start of World War II. The book is bitter, a kind of nightmare about tyranny and bullies. In it, the main character George

Bowling faces his fear of financial insecurity, a life of endless work, an empty marriage, and the impending war. Bowling's escape is fishing and the nostalgia he has for fishing as a child. Embedded in Bowling's ecumenical fear is a yet-to-be articulated concern for the condition of the world not merely politically, but in what we might now identify as an environmental anxiety, a hint of ecological thinking born of his desire to go fishing. Orwell offers a number of magnificent insights about fishing through Bowling. More than once when wading through policy debates, I have found myself thinking of Bowling's words "fishing is the opposite of war" and rebutting "no, George; fishing has become, in fact, a rather hostile war." Or, as Jeff Angers, my good friend and president of the Center for Sportfishing Policy, puts it, "fishing is not supposed to be political, but the reality is, quite simply, that it is political."

Of course, fishing has caused political hostility since long before Orwell took on the subject. We can easily trace fishing to the center of New World exploration and colonization and to the heart of US history right up until this very moment. It is at this moment, however, that more than eleven million recreational saltwater anglers have the opportunity to reimagine and reinvent what it means to live the saltwater fishing life and what the responsibilities of that life stimulate. Not to return to an imagined golden oceanic heyday, but to reimagine under the shadow of overfishing, climate change, and pollution, to reinvent in a way that approaches the neoteric to account for contemporary human interaction with fish and ocean. This, then, is a book about the fear of regretting what has not yet happened—and how to imagine ways to preempt that regret.

Deliberating about policies that address marine environments and fisheries management will always be problematic, not simply because they pull at presumed cultural and economic entitlements, but because they can only ever be addressed within a covenant of immediacy. Despite any nod to policy development that accounts for future or past thinking about "resources," such policies are inevitably formulated within a deep-rooted concern for the present, within a mind-set that formalizes fisheries populations as, most importantly, quantifiable and policy outcomes as valuable when they provide economic benefit within extant perceptions. Ideologies of immediacy require significant reconsideration. Such deliberations will always be charged, and perhaps dangerous, but are also indispensable.

Thus, as recreational anglers we find ourselves at a moment that requires recognition of such dangers. Avoiding them, pretending they do not exist, will only make the act of deliberation more difficult. Instead, we must engage the difficult problems. To do so, we must extend our thinking beyond the immediate and beyond proximate futures to a more sustainable understanding of recreational fishing's distal futures, possible futures situated inherently away from a specific point of origin or presumed outcome. If we consider proximate futures, our immediate, impending futures, we can identify the preceding causes and effects that are likely to lead to that future, a chain of events such as those I have described regarding policy. Thinking, though, with distal futures in mind provides longer-term vision wherein causes are less immediately identifiable, as in long-term evolutionary processes. We can build toward planned proximate futures, but the ability to think in distal futures will be crucial. Our proximal futures are already here. Long-term sustainable thinking requires a more complex ethical obligation to extend our willingness to theorize about distal futures.

This book uses conceptual approaches to think about the distal futures of recreational fishing. Because it has been dubbed the "most contemplative of pastimes," recreational fishing provides a valuable perspective on how humans interact with saltwater environments. Each of the concepts taken up in these chapters is derived from academic work I have pursued over the last twenty years of research about writing, ecology, and technology; yet this is no academic narrative. Instead, *Fishing, Gone?* is speculation. It is an often-uncomfortable interrogation of the saltwater sportfishing life as a method for more carefully thinking through the current status of saltwater fisheries and the environments they inhabit. This is a book of fishing stories. It is a consideration of what it means to tell fishing stories and what those stories might offer us as points of departure for thinking about marine environments. It does not strive to be prophetic, but to envision fishing yet to come and stories hidden deep, still to be told.

In the preface to his powerful examination of the history of industrial fishing, *The Unnatural History of the Sea*, Callum Roberts explains that his book "is not a requiem for the sea," and that he maintains hope for the future of the ocean. He explains, "We still have time to reinvent the way we manage fisheries and protect life in the oceans." He is, he writes, "optimistic for the future."[28] I want to share Roberts's optimism, but until we witness drastic reinvention not only of global fisheries policies but also of deep-seated cultural presumptions about fish, ocean, and life more generally, hope remains elusive for me. In the pages that follow, I consider how such reinventions might unfold in a new oceanic imagination for the recreational angler. This is a reflection upon the water. It is a suspended moment of curiosity and concern. It is recognition that those of us who fish the ocean can no longer live for the thrill of the strike, for the instant of exhilaration. It is not about living for the day on the water or for the escape of fishing, but about living for the ocean, with the ocean, and through the ocean. It is about the opportunity to fish on.

2

The Ecstasy of Space

The only other place comparable to those marvelous nether regions, must surely be naked space itself, out far beyond atmosphere, between the stars, where sunlight has no grip upon the dust and rubbish of planetary air, where the blackness of space, the shining planets, comets, suns, and stars must really be closely akin to the world of life as it appears to the eyes of an awed human being in the open ocean, one half mile down.

William Beebe, *Half Mile Down*

It was hot—deep-summer Everglades hot. The kind of hot to which Hell aspires. Roy and I had been fishing the inshore waters of Ten Thousand Islands for a few days. These waters, on the southwest coast of Florida, in the northernmost section of Everglades National Park, are legendary for their fishing diversity. A giant nursery for marine life—more than one million acres of protected waters—the area is a giant buffet for sport fish: cobia, grouper, kingfish, mackerel, permit, pompano, redfish, shark, sheepshead, snapper, snook, tarpon, triple tail, and trout, as well as juvenile jewfish.[1] We had been fishing, but we had not been catching, and that was starting to frustrate us both—that and the sticky swelter. There was no breeze; the water was as flat as flounder, as Beaumont and Fletcher's hungry servant Penurio would have put it. The no-see-ums and gnats in the mangroves certainly did not help. "Screw this," we agreed. "Let's go offshore."

We were probably as eager for the flow of air from the boat running across the slick surface as we were for the prospect of offshore fishing. We ran hard for an hour or more, hammer down, gliding on that flat, open, green water, putting lots of miles between us and the shore. The air roared in our ears, offering a bit of relief from the heat. I pulled back a bit on the throttle so we could talk about what we should do. The boat settled from its plane atop the water into a comfortable midhull float. The depth finder showed depths ranging from 100 to 150 fathoms, deep water for the Gulf. Roy looked at those numbers. "I'm getting in," he said. "I gotta know what it feels like to swim in a thousand feet of water."

"It feels like swimming in six feet of water or sixty feet of water," I replied. "It feels the same because you stay in the same place at the surface."

"Yeah," he agreed, "but I mean what it feels like to know there's that much water under you, that it's a hell of a long way down there, and that you have no clue what else is down there."

I dove from the platform where I was sitting before he finished explaining. I had to know that, too, and in that instant between boat and water, I struggled between the desire to know and an innate human thalassophobia. In that breath between surrendering contact with the boat and breaking the water's surface, in that moment of free flight in the air, the ecstasy of the sublime washed over me. It would have mattered little if I had fallen two hundred fathoms instead of ten feet. I was moving through space.

This is the oceanic sublime—the ecstasy of inconceivable fluid space, the desire to comprehend and manage, and the inability to do so. To fish in saltwater is to reconcile one's insignificance in size and space. To pursue fish in saltwater is to hunt in a vastness under inherent conditions of limit imposed by space and physical property. For the angler, that enormity tempers all possible thinking about fish and fishing. There can be nothing more sacred to the saltwater angler than the space of sea.

Earth is a planet of saltwater. Conventional accounts simplify it: saltwater covers nearly 71 percent of the Earth's surface. That translates to roughly 128 million square miles of ocean surface and 310 million cubic miles of ocean volume. Relatively speaking, that is a hell of a lot of saltwater (and just about anyone writing about marine environments is apt to relate that fact, as though the relative amount of saltwater stands as an argument in and of itself).

There's a scene in Michael Bay's 1998 blockbuster movie *Armageddon* in which the president of the United States (played by Stanley Anderson) asks NASA's Dan Truman (Billy Bob Thornton) why NASA didn't know sooner about the planet-killing asteroid headed for Earth. "Well, our object collision budget's about a million dollars a year," Truman replies. "That allows us to track about three percent of the sky, and begging your pardon, sir, but it's a big-ass sky." We like to think that it is also a big-ass ocean.

The average depth of the ocean is about 12,080 feet, and the deepest point is Challenger Deep, which reaches 36,200 feet (just under seven miles) at the southern end of the Mariana Trench in the Pacific. Humans have seen less than 5 percent of this vastness—and less than 1 percent below one thousand feet.

Think about it this way: the average US master bedroom measures just over two hundred square feet. If we understand that room as having a standard eight-foot ceiling height, we can consider it to have a volume of sixteen hundred cubic feet. Imagine, then, peering into this room, knowing how big it is but being able to see a cube of only two feet by two feet by four feet—a space about the same size as six standard cinder blocks stacked together. The rest of the room is there—you know it is there—but you can see only the space the cinder blocks occupy. The rest of the space is unknown, unseen. You can only assume what the space is like and what occupies it by imagining that it is remarkably similar to that small portion you can see. Now contemplate living in that room, seeing only that 1 percent view. The remaining 99 percent may seem vast, empty, treacherous, or even inhabited by monsters, all depending on the myths, legends, and assumptions made about that space over time.

The myth of the ocean's limitlessness, a myth we have willingly accepted, willingly imagined as akin to the myth of outer space's perceived (or at least assumed under our current understanding) endlessness, is being not only disproven but revealed as an illusion conjured in ignorance. The ocean, we must acknowledge, is neither boundless nor immense. Relatively speaking, yes, it is enormous, but it is confined, contained, and finite. Only with a grasp of limit can we begin to rethink the relationships between ocean, fish, and angler in productive ways. The ecstasy of boundless space has driven not only the ways in which we perceive the oceanic, but the very ways in which we confine our abilities to think about oceanic space (and, perhaps by association, space in general). It might better serve the ocean—and, by extension, us—to adopt a philosophy of limit in how we think about, talk about, write about, manage, and imagine space. Oceanic space reasserts itself in a sublimity of encapsulated size, its vastness recontextualizing our knowledge and beliefs. We can no longer think of ocean in terms of inexhaustibility, and it is unlikely that any serious-minded conservationist-angler still does. Space, that is, is not synonymous with endless.

Within the narrative that taught us to believe in the ocean's boundlessness, we learned to believe, too, that what dwells within the ocean must also be limitless. Of course, the prevailing state of mind about the limitless ocean results from a view limited by technological ability; if we cannot see the end, we assume there is no end. As humans, we require complex technologies to see the limits of the ocean. Our individual perceptions confine our ability to think within limits. Only through technological augmentation—boats, satellites, drones, sonar, scuba, deepwater submersibles, and the like—do we begin to identify and map the very limits we have assumed to be beyond our knowledge. The value of the ocean lies not in its vastness, but within its limits. Those limits must now guide our angler philosophies, lives, and actions—not just on the water, not just when we are fishing, but at all moments in all times. For the angler, this requires embracing ways of thinking inspired by fluidity.

I am reminded of Scottish author R. M. Ballantyne's 1874 book *The Ocean and Its Wonders*, written just twenty years after Commander M. F. Maury described the Gulf Stream current and fifty-four years before Cousteau or Beebe would write of their descents into the ocean. Ballantyne wrote in an era of relatively profound oceanic unknowing, conceding repeatedly to scientific ignorance about the ocean, including what now seems commonplace knowledge such as the depth of the ocean. Ballantyne attests to the technological inability to measure depth, recounting efforts using weights and lines of various compositions. He explains that the ocean's "depth has never been certainly ascertained" and probably never will be. By his account, conventional wisdom confirms the likelihood that the ocean could never exceed a depth of five miles. Yet despite Ballantyne's concession to ignorance, he is enamored, caught in the oceanic sublimity. He gracefully writes:

> There is a voice in the waters of the great sea. It calls to
> man continually. Sometimes it thunders in the tempest,
> when the waves leap high and strong and the wild winds
> shriek and roar, as if to force our attention. Sometimes it
> whispers in the calm, and comes rippling on the shingly
> beach in a still, small voice, as if to solicit our regard. But
> whether that voice of ocean comes in crashing billows or
> in gentle murmurs, it has but one tale to tell,—it speaks of
> the love, and power, and majesty of Him who rides upon
> the storm, and rules the wave.[2]

In the instant before I broke the surface of the Gulf that hot day,
I looked west toward open water, seeing before me horizon and
a vast surface where air and water met. As I dove through air,
breaking the surface of the water with hands and head leading, for the briefest of moments, a mere fraction of a second,
I inhabited both air and water, crossing from troposphere to
hydrosphere, enveloped in the very biomedia that permit and
encourage life to flourish on this planet. In that instant of being
neither here nor there, being wet and dry, being wrapped in
gas and liquid, I surrendered to the world of fluidity. Gravity
pushing and pulling me down through the air, the water buoying
me up. Simultaneously falling and floating. So fleeting the
moment—a blink, if that. In a flash, I submerged with fathoms
below and a few feet of saltwater above. Here is where the angler
lives at all moments of life, caught between air and water and
embraced in their limits.

Fluidity and limits; these are complex ideas that require
anglers to afford them more than cursory consideration. In
2010, following the Deepwater Horizon oil rig disaster in
the Gulf of Mexico, I read National Geographic explorer-in-
residence Sylvia Earle's *The World Is Blue: How Our Fate and*

the Ocean's Are One with a good deal of enthusiasm. In those pages she shows just how historically entrenched the ideology of limitless ocean is, citing an 1883 Thomas Huxley speech in which he famously said:

> The herring fishery, the pilchard fishery, the mackerel fishery, and probably all great sea fisheries are inexhaustible: that is to say that nothing we do seriously affects the number of fish. And any attempt to regulate these fisheries seems consequently . . . to be useless.[3]

This same attitude of inexhaustibility drove unbridled economies of cod, mackerel, herring, and menhaden harvests starting as early as the thirteenth century. While the economic significance of large-scale fish harvesting remains central to any conversation regarding the future of the world's ocean and fisheries management, we must also account for more than nine hundred years of cultural accordance and disposition that reduce fish—all fish—to economic artifact rather than living organism emphatically bound to global ecosystems and human lives. Even as we begin to understand the necessity of seeing fish as an essential node in complex networks of oceanic and terrestrial ecosystems, we do so much more apathetically than we do with terrestrial organisms or oceanic mammals. If we are to move toward practices that promote ocean sustainability, then we must first move toward ways of thinking about fish and fisheries beyond mere economic frames.

Take, for instance, the ongoing situation regarding management of the Gulf of Mexico red snapper fishery, one of the most problematic and contested fisheries management issues in US history. To oversimplify the situation, the Gulf of Mexico Fishery Management Council—one of the eight fishery management councils organized within the National Marine Fisheries Service, a division of NOAA—uses a portion allocation system in determining how many red snapper recreational anglers can harvest and how many industrial harvesters can take. Until very recently, the commercial and industrial sector has claimed a larger portion of the take, with an allocation of 51 percent for commercial take and 49 percent for recreational take. In 2016, an amendment to the Fishery Management Plan for the Reef Fish Resources of the Gulf of Mexico was proposed to reallocate the portion division to 48.5 percent for industrial and commercial harvest and 51.5 percent for recreational harvest. Whether this switch indicates a larger shift in the economic impacts of recreational versus industrial markets is something to be taken up later. However, in terms of rethinking how we consider the limits of fisheries populations within a history molded in ideologies of limitlessness, what is significant, first, about these allocations is that the council apportions them in weight, not numbers of individual fish. Measuring in this way presumes a correlation between a fish's size, its maturity, its capacity for breeding once mature, and, ultimately, not the value of the individual organism, but that organism's contribution to a total mass. We can think of this as tonnage-thinking.

Tonnage-thinking contributes to a cultural understanding of the red snapper's value occurring in weight, not in the fish itself. The red snapper, for example, is commonly known as one of the most valuable fish in the Gulf of Mexico—a value expressed only in economic terms. Nevertheless, when we attribute the effects of tonnage-thinking to red snapper, we can begin to better understand how and why the economic value dominates how we think of the fish's role in the ecology of the Gulf of Mexico. Red snapper, we know, can live to be more than fifty years old. They can reach about forty inches in length and can weigh up to fifty pounds each—*can weigh*—this is now a rarity. Red snapper begin to spawn when they are about two years old, and larger red snapper produce more eggs than smaller red snapper. Estimates show that a single eight-year-old red snapper can produce as many offspring as 212 five-year-old snapper. However, decades of overharvesting have reduced not only the Gulf of Mexico's stock of red snapper but also the age of the stock. Current research shows that most of the Gulf's red snapper are between four and six years old. The removal of older, larger fish has reduced the rate of reproduction of the entire population, as the population depends on younger fish to provide nearly all stock reproduction. Because younger fish produce fewer young than do older fish, the harvest of the larger fish—those of greater weight and age—sets the overall population back in reproduction rates. Likewise, if larger fish are harvested as a more efficient method for achieving stock harvesting allocations—that is, achieving weight allocations— the focus on total weight will always cull from the breeding end of the population. Tonnage-thinking might be functional for an industry looking to most efficiently harvest its allocation in the most financially feasible way, but the measure of three three-pound fish versus one nine-pound fish cannot be reduced to simple pound-for-pound thinking. To reconceptualize what it means to harvest, we must first set aside tonnage-thinking as the measure by which we imagine the value of fish.

I should note here, too, that tonnage-thinking is not tied merely to how we imagine commercial harvest; it is deeply rooted in recreational fishing, too, as we often attribute value to a catch in the weight of the fish. Choy's Monster, for example, the legendary 1,805-pound marlin caught in 1970 off Oahu, earned its status not from the gallantry of the battle in taking it, but strictly from its weight. The same is true of Alfred Glassell's 1,560-pound black marlin, caught in 1953, or even Craig Carson's 17-pound, 7-ounce speckled trout, caught in 1995. While this logic of size has existed in the angling mind-set for hundreds of years, I find it problematic because the weight of a fish is of value only once it is removed from the water. The buoyancy of the ocean and the physiology of the fish allow the fish to adjust how its weight interacts with the ambient environment. A five-pound speckled trout may feel heavier during an angling contest than a seventeen-pound trout, depending on many characteristics beyond simple pound-for-pound weight.

We have made a dubious exchange with the ocean, extracting and slaughtering immeasurable quantities of its inhabitants and reciprocally depositing unbounded sums of waste in it. In the last seventy-five years, humans have removed as much as 90 percent of the populations of many fish species. The human demand for the meat of some populations, such as the Atlantic cod and the bluefin tuna, has reduced those species by 95 percent. Yet it is not just the harvesting of fish per se that contributes to the overall degradation of the world's ocean. The technologies and techniques many commercial and industrial fishermen use not only remove targeted species in bulk but are responsible for near-endless bycatch of other fish and marine animals, including marine mammals, birds, turtles, and invertebrates. Likewise, many of these same technologies and techniques actively destroy marine habitats through indiscriminate harvest practices. Dredging and trawling, for example, strip ocean floors of animal life, plant life, and topography. Deepwater reefs, which for years were beyond the reach of humans, are now being crushed and stripped away by trawling and dredging. In the last seventy-five years, nearly 50 percent of the world's shallow-water reefs have been destroyed, and the slow growth rate of coral means that rebuilding long-term colonies could take centuries. The marriage between technology and culture can be conveniently vilified as culpable in the destruction of the world's ocean. However, it is this same cultural-technological tandem that stands at the center of our ability to think ahead to what the world's ocean might become and how we might begin to put in place new ways of thinking, new philosophies, and new practices that will guide how we see and use marine space. That is, we cannot blame technology itself for the ways in which various technologies are used destructively, when we also depend on technology to provide solutions to our oceanic demise. In fact, it is specifically cutting-edge technologies on which we must rely to mold the ocean of the future.

Think about this: more than half of the global population inhabits coastal regions; that's about 3.5 billion people. Estimates show that by 2025, 75 percent of the world's population will live on the coast. In the United States, more than 80 percent of the population lives within an hour's drive of the Great Lakes, the Atlantic Ocean, the Pacific Ocean, or the Gulf of Mexico. The populations of these coastal regions continue to grow; some research shows that approximately 3,600 people move to coastal regions *each day* in the United States.[4] Population increase, of course, compounds a number of complications for ocean environments: increased runoff and pollution from industry and sewage, increased offshore sewage disposal, landfill of wetlands to support human development, increased importation and eventual runoff of freshwater, and increased emissions, to name a few. We might categorize these as coastal problems, but their effects circulate throughout the ocean's currents. We might also characterize these problems as technological issues, but as I have said, such vilification is naive, as condemning technology writ large as causal is both fallacious and heedless. Moreover, as I hope to show, the future of the world's ocean and its nonhuman populations may in fact be utterly dependent on human technologies to facilitate and maintain their vitality.

Of course, making such a claim is deeply problematic for a number of reasons; the most evident is that the ocean will continue to exist in some form no matter the pretension of human intervention. Certainly, the risk is that the ocean may not continue as we imagine it should, but it will continue in one form or another. Our oceanic imaginary arises from the same human-centered ego that validates our destruction of the ocean, because we inevitably tell ourselves that we know what is right. Nature, however, does not give a shit what we know.

The vastness of the ocean and the enormity of the destruction wreaked thus far can be more than difficult to fathom, let alone comprehend in any meaningful way. Think about this: Earth is about 4.6 billion years old. To make that more manageable, scale that to forty-six years. Comparatively, within this scale, humans emerged about four hours ago. The industrial revolution cropped up just about one minute ago. In that same minute, we have managed to destroy more than 90 percent of the populations of many fish species. Unquestionably, the relationships between industrial technologies and cultural values have come to clash with ocean and fish in material and philosophical ways. Our thinking must take on a new logic, a divergent approach to our cultural imagination about oceanic vastness and inexhaustibility. The ocean is big, yes, but not limitless.[5] We now test those limits and must assent to their boundaries, no matter their overwhelming size. To do so requires thinking not about what we once believed the ocean was in size and resource, but about what it might become.

Shifting our thinking in this way is not a matter of striving to return the ocean and fisheries to what they once were. The present is always unique, not analogous with any past, if for no other reason (though there are many) than the simple fact that humans and human technologies did not occupy as much of the past as they do the present. Philosophies of recoupment or reclamation are shortsighted, assuming that one part of a complex system can be returned to previous conditions without returning all of that system to the conditions that fostered those circumstances sought from reclamation. We cannot return fisheries populations to the numbers they once were without returning the size of the human population to what it was at that same historical moment. Or returning to the same technological capacity we had when fish populations were what we wish they now were. Or returning the economic value of those fisheries to what it once was—a change that would perhaps be met with more resistance than any other. Those moments have passed. Like the fish lost or released, we can wax nostalgic about those times, but it is time to plan the next cast.

We must begin thinking remotely in order to account for the economic and political power of recreational saltwater angling in ways that were previously not possible and that have not accounted for that population's generally encompassing conservation ethic. We must think differently about the ocean and its inhabitants.

The ocean is liminal space, ensnared between a cultural history of imaginative unknowing and a technological revelation that exposes only hints of what is yet to be known. We know that we do not know a lot about the ocean, but we also know that we will know more—perhaps rapidly. What we do with our knowledge yet to come will be critical to the ocean's future. Part of such radical thinking will require that we ditch the foundational

perception of nature as a constant; we must reimagine nature—
and thereby ocean—as being always already in nascent states
of natural. Think of it this way: at this moment you are a much
different person than you were a year after you were born. And
while you retain some elements of your younger self, you can
never return to the exact characteristics and constitution of that
self—no matter how badly some may want to recapture physical
or emotional aspects of the past—because too many external
factors and physiological changes have rendered that version of
you past tense, yet still you. The same may be said of the ocean,
and no matter our attempts to enact pollution control, curb overfishing, or restore the ocean in other ways, restoration will never
be possible, simply because the ocean is connected to too many
complex ecological parts that would also have to be restored to
earlier incarnations. Short of Mr. Peabody's Wayback Machine,
these are reclamations beyond our grasp.

As liminal space, the ocean is a space of transition and emergence; it is a space on the threshold. It is a space encumbered by
time and the ever-asked question "what's next?" The ocean is ever
changing, always on the cusp not only of what is next, but also
of a new beginning, often dependent on disruption to initiate
necessary changes. Just as a hurricane does not wreak destruction
on a place, as humans tend to mythologize, but causes natural
alterations to changing spaces of water, seabed, and shore, so
too do disruptions alter the next incarnation of liminal space.
Anglers—and humans in general—have opportunities to enact
cultural disruptions that can affect liminal space, both posi-
tively and negatively. My hope is to be productively disruptive,
to suggest other ways of engaging ocean to the end of making
its future. Is such intent presumptuous? Does it reek of a god
complex and homocentric control? Perhaps. But to imagine that
humans have not previously exerted such presumption over this
planet and its inhabitants would be a denial of our humanness.

I turn to Rachel Carson in *The Sea around Us*, to her contemplations about the human relationship to ocean:

> Eventually, too, man found his way back to the sea. Standing on its shores, he must have looked out upon it with wonder and curiosity, compounded with an unconscious recognition of his lineage. He could not physically re-enter the ocean as the seals and whales had done. But over the centuries, with all the skill and ingenuity and reasoning powers of his mind, he sought to explore and investigate even its most remote parts, so that he might re-enter it mentally and imaginatively.[6]

But Carson missed the boat, as it were, on what that imagination would render. "And yet he has returned to his mother sea only on her own terms. He cannot control or change the ocean as, in his brief tenancy of earth, he has subdued and plundered the continents."[7] Of course, we have returned to the sea—or more accurately, have never left it—violating her and her terms in countless ways, changing her beyond return.

In 2003 the United Nations' first *Global Environment Outlook Yearbook* identified 146 hypoxic zones throughout the world's ocean; the rapid increase in numbers is easily attributable, but difficult to curb. Commonly known as "dead zones," hypoxic zones are ranges of ocean space with such low oxygen concentration that no animal life can survive there. Some hypoxic zones occur naturally, but most are human induced through biological and chemical means. Most anthropogenic hypoxic zones result from nutrient pollution, a condition in which nutrients from fertilizers wash into rivers and streams and eventually into the ocean, stimulating algal growth. Decomposing algae fall to lower depths in the water, where decomposition depletes the water of oxygen. More chemicals lead to more algae; more dead algae lead to more oxygen consumption in decomposition; and more oxygen consumption results in reduced levels of oxygen in a region, or a "dead zone."

One of the largest of the world's hypoxic zones appears every spring in the northern Gulf of Mexico as fertilizer and chemical runoff from farms along the Mississippi River spills into the Gulf, choking out the oxygen and in turn suffocating or chasing away all animal life in the region. This kind of hypoxic zone is the result of complex relationships between industrial farming and marine environments—relationships that need to be considered not just ecologically but politically and philosophically, as well. Imagine the impossibility and senselessness of arguing that we should cease farming or using fertilizers along the Mississippi and throughout the Midwest as the corrective for Gulf hypoxia. Or more dramatically, imagine succeeding at such an argument and the national and global results of eliminating the core of American agriculture. These are not the kinds of solutions needed.

In 2008, the journal *Science* identified more than four hundred hypoxic zones globally, more than double the number counted in the 1960s. According to the report, the hypoxic zones affect more than 152,000 square miles of ocean and should be considered one of the primary stressors of marine ecosystems.[8] Because hypoxia is also an anticipated outcome of long-term climate change, scientists have been able to measure analogous historical scenarios to estimate recovery times from long-duration hypoxia. Examining mollusk fossils dated from 3,400 to more than 16,000 years ago, researchers from the University of California, Davis have been able to show that recovery times from even short bouts of seafloor hypoxia can take around one thousand years.[9] Certainly, in the larger chronological scope, a one-thousand-year recovery time may seem only a blip in the global ocean's life, but when compounded across a seemingly ever-increasing space of hypoxic zones, recovery becomes more of a challenge and the once-measureless ocean begins to appear segmented, broken, and limited. Our boundless ocean bound and constricted by hypoxia.

On July 8, 2011, at 11:29 a.m., the 135th and final space shuttle flight took off from the National Aeronautics and Space Administration (NASA) facility at Cape Canaveral, Florida. Launching from Pad 39A, STS-135 concluded a nearly forty-year program designed to support construction of the International Space Station (ISS) and deliver payloads into low Earth orbits. This final mission of the orbiter *Atlantis* carried commander Christopher Ferguson, pilot Doug Hurley, mission specialist Sandy Magnus, and mission specialist Rex Walheim to the ISS, where they delivered a Raffaello multipurpose logistics module. At 11:29 that summer morning, I stood with my wife and sons on our deck 130 miles away from Pad 39A and clearly saw the solid-fuel rocket boosters fire and the iconic white contrail of that historic flight.

Since 1981 I have watched dozens of launches blast off from the NASA pads: shuttles, Titans, Atlases, Deltas, and Saturns; sometimes I've stood as close as ten miles away from the launchpads. From various locations along the southern waters of Florida's Mosquito Lagoon, NASA launchpads are clearly visible. During the years of NASA's space shuttle program (1981–2011), it was not uncommon to look up from the water of the lagoon to see a shuttle positioned at a launchpad awaiting its next mission. Pads 39A and 39B are located a few hundred feet west of the Atlantic Ocean and less than seven miles south of Mosquito Lagoon, one of my most frequented fishing sites.[10] It is impossible, in fact, to fish the southern reaches of

Mosquito Lagoon without feeling the ubiquity of NASA or the ambient presence of space that NASA projects.

Mosquito Lagoon is one of three lagoons that make up the Indian River Lagoon (IRL) system along Florida's northeast coast. The IRL[11] extends 156 miles from the northern point of Ponce de Leon Inlet in Volusia County to its southern point at Jupiter Inlet in Palm Beach County. Together with the Banana, Indian, Eau Gallie, and St. Sebastian Rivers, as well as Hobe Sound and Haulover Canal, the IRL is the most diverse estuary in North America, housing more than 2,100 species of plants and 2,200 species of animals, 35 of which are threatened or endangered, more than in any estuary in North America. Spanish explorers named the IRL Río de Ais after the native population of Ais Indians. The derivative "Indian River" still names part of the IRL system.

I have fished the IRL since I was a boy, particularly Mosquito Lagoon. In fact, fishing the lagoon from my kayak over the last dozen or so years has become some of my favorite fishing. Mosquito Lagoon is legendary in the world of inshore fishing. It boasts the monikers "Saltwater Shallow-Water Fishing Capital of the World" and "Redfish Capital of the World," though the area south of Hatteras, North Carolina, between Avon and Buxton may contest the latter given that David Deuel's 94-pound, 2-ounce all-tackle world-record redfish was taken from the beach at Avon in 1984—a record that still stands. Nonetheless, Mosquito Lagoon has produced many IGFA (International Game Fish Association) world-record redfish and spotted sea trout (what we call "gator trout").

Unlike the shores of the rest of the IRL, Mosquito Lagoon's shores are free of any residential development; there are no businesses or commercial property anywhere along the 2,100 acres and twenty miles of lagoon shore—a rarity, to say the least, in Florida. There is no direct agricultural runoff into any part of the lagoon. No causeway or bridge cuts across it; the barrier island that frames the east side of the lagoon is accessible from the mainland only north or south of the lagoon. Its waters are as pristine as one can find anywhere in Florida. The lagoon's bottom is a giant grass flat, spotted with sand potholes and oyster bars, ideal habitat for redfish, trout, and other inshore species. Water depths average less than three feet throughout the lagoon, and in many places the water is never more than a few inches deep. The shallow, (usually) clear water provides excellent conditions for sight casting to remarkably large fish. Most of the lagoon is part of the Merritt Island National Wildlife Refuge, its management overseen by the US Fish and Wildlife Service.

However, Mosquito Lagoon and the rest of the IRL are at serious risk of hypoxic destruction. By 2013, the IRL's reputation as the most productive US inshore fishing location was being overtaken by its reputation for massive algal blooms and fish kills. In a 2013 article in *Wired* magazine, Brian Lapointe, a marine environmental scientist from the Harbor Branch Oceanographic Institute at Florida Atlantic University, put it bluntly: "The Indian River Lagoon has become a toilet."[12] Lapointe's metaphor proved to be more accurate than one might hope when in 2016 the IRL became the site of two of the worst hypoxic-based flushes in human history. In March 2016, a bloom of *Aureoumbra lagunensis*, commonly known as "brown tide," killed thousands of fish in the IRL. This bloom resulted from decades of runoff composed of septic drainage, fertilizers, and stormwater. To make matters worse—and certainly more confusing to local anglers and residents—at this same moment, large quantities of polluted andnutrient-rich water from Lake Okeechobee and the St. Lucie basin were released into tidal waters, contaminating not only the IRL but Florida's southwestern waterways, as well. In its natural state, the Okeechobee runoff had flowed slowly through South Florida and the Everglades into Florida Bay. In its natural state, too, this water had very low levels of contaminants such as phosphorus and nitrogen. Natural flow of this water could take six months to circulate from lake to bay. However, human intervention rerouted the natural flow to the east and west of the lake, and controlled flood management forced the flow to move much more rapidly—about six times as fast. Likewise, the "managed" waters now contained tremendous amounts of toxins, most of which could be attributed to runoff from sugar farms, as the result of shady relationships between Florida's politicians and the sugar industry—what we call Big Sugar. In February 2016, the sudden rush of toxic water into the IRL and the Caloosahatchee watershed turned the waters black, choking out fish, marine mammals, shellfish, and birds. Images of fish kills and unnaturally black water dominated local fishing conversations and national conversations about pollution, environmental degradation, and politics. Voters petitioned Governor Rick Scott to close the dams that had been opened to allow the runoff. In February 2016, Governor Scott declared Florida's Lee, Martin, and St. Lucie Counties disaster areas. Scott blamed President Obama for failure to fund $800 million to repair the Herbert Hoover Dike, which was built around Lake Okeechobee in the 1930s for flood control.

But Florida's native son and vocal defender Carl Hiaasen quickly pointed out that Scott clearly did not understand federal appropriations procedures and, as Hiaasen was apt to do, pointed out that Scott's impending run for Congress was being funded by the very industry leaders in Big Sugar that had polluted the waters across Florida.[13] Hiaasen also noted that Scott's attempt to blame the sitting Democratic president failed to account for the role Scott's own Republican party had played in Congress failing to move forward with funding to repair the dike. Soon after Hiaasen's criticism of Scott's handling of the runoff, *Huffington Post* blogger Alan Farago laid the blame for the IRL and Caloosahatchee situation on Florida voters:

> Blame the Florida legislature, its leaders like Agriculture Secretary Adam Putnam, Florida Representative Matt Caldwell, House Speaker Steve Crisafulli, Senator Joe Negron, to start. Blame the most clueless, radical governor in Florida history—the anti-people governor Rick Scott who never addressed a problem that he couldn't visualize from 30,000 feet in his private jet, or, if he did then just slammed the window shades shut.
>
> There is no need for science to tell us what's wrong with Florida's waterways and the cascade of destruction from Florida Bay stretching north along both coasts. We know exactly what is wrong: voters who don't care, don't vote, or vote for candidates and incumbents who represent institutionalized corruption.[14]

In April 2016, I sat atop my kayak in the northernmost stretches of Mosquito Lagoon's maze of mangroves and skinny water, the New Smyrna skyline visible to the north. The water was muddied by the wind, but it was the cleanest water I could find. The southern reaches of Mos Lag were a mess. For two months I had been hearing conversations in tackle shops from Titusville to St. Lucie about the runoff, how miserable inshore fishing was, the direct effect on local economies, the reduced numbers of tourists, the guides without charters, and, mostly, the dead fish. Sitting there, I cast an artificial shrimp—a lure made from a new, high-tech thermoplastic elastomer (TPE plastic), the newest rage to hit the lure industry. The thing looked more perfect than the most perfect shrimp. It had been designed using high-resolution, 3-D scans of actual shrimp and was virtually indestructible. It was beautiful, a work of art. Hypershrimp. More shrimp than shrimp. Shrimp plus. The ubershrimp. I coated it in a manufactured gel designed to make the artificial shrimp smell like a real shrimp. I adored this shrimp the same way I might a forty-three-foot Viking Convertible or a '66 Corvette. Things of beauty.

My cast was soft and subtle, directed to the left of two redfish tailing in the mud next to a mangrove shore. One tail went down and my line went tight. The fight was solid, but short. A slot-sized red. Low end of the slot, but still a legal-to-keep fish. I considered the moment: a high-tech artificial shrimp in the northern reaches of the infamous Mosquito Lagoon and Indian River Lagoon, with its ill waters south of me in the shadow of NASA exploration. In that moment I meditated on the one ocean, lost in its vastness and vexed by its finitude. There, cradled in water and space, I slid that fish back into the lagoon.

3

Alien Ichthyology

> If some alien called me up . . . "Hello, this is Alpha,
> and we want to know what kind of life you have,"—
> I'd say waterbased. . . . Earth organisms figure out
> how to make do without almost anything else. The
> single nonnegotiable thing life requires is water.
> —Christopher McKay, NASA scientist, *Omni*, July 1992[1]

Since I can find no such book, I often imagine writing a
particular kind of book about fishing. This imagined book
would not be one about the fish we find in our ocean, but a
book—perhaps science fiction—about how we might identify
a fish should we find one on another planet, in another galaxy,
or even one here we don't recognize as a fish as such. That is, I
want to write a book about alien ichthyology, not about fish per
se but about the very idea of what it might mean to be a fish.

Of course, the desire in writing such a book is speculative,
perhaps taxonomic, but it is a response, too, to an ancient
desire to classify living things, to place organisms in a natural
order, whether phylogenetic or biblical. The speculation in such
a book would extend from the physical characteristics of what
constitutes a fish to environmental and ecological questions
concerning what the habitat of a fish might be and how the
identification of fish relies as much on the habitat in which it
lives as on its own body. Is a fish out of water still a fish? Such
questions might be as likely to originate from exobiology as from

ichthyology. Terran ichthyologists would rely on ambiguous definitions of "fish" such as "a poikilothermic, aquatic chordate with appendages (when present) developed as fins, whose chief respiratory organs are gills and whose body is usually covered with scales,"[2] a definition crafted toward diversity and possibility. Some definitions, such as Joseph S. Nelson's, stretch the ambiguity even further in their simplicity: "Despite their diversity, fishes can be simply, but artificially, defined as aquatic vertebrates that have gills throughout life and limbs, if any, in the shape of fins."[3] Biologists recognize the term "fish" to suggest not so much a definition as a general description for a diverse array of aquatic organisms. However, the question "what is a fish?" is as much cultural as it is biological or ichthyological, and for those of us who float about in the culture of saltwater recreational fishing, the question has deep ramifications.

Most of the world's animal organisms live in water. In fact, if McKay's Alpha were to arrive on Earth, it is unlikely that it would identify land or land-based animals as the focal point of the planet's life systems. Fifty-eight percent of the world's fish species live in saltwater—approximately fourteen thousand species. The International Game Fish Association (IGFA) identifies eighty-three of these as saltwater game fish, but recreational anglers worldwide target significantly more of the fourteen thousand species, particularly those of higher food value even if they have little or no sport value. It is precisely these values and value systems that we must now rethink. Historically, we determined the cultural value of fish by economic value, primarily in terms of food harvesting, whether individual or industrial, and other manufacturing value, such as oil extraction. Much later, and to a lesser degree, recreational harvesting introduced a new value tied to sport and leisure. These values, of course, must change as the planet's ocean is remade in contemporary context. The value of fish as critical components of larger, complex ecological and biological systems must emerge as central to the remaking of the world's ocean. Recreational anglers are positioned at the heart of this cultural shift because such changes are not merely scientifically vested in terms of ecology and biology, or culturally vested through economic values, but epistemologically necessary in building a new worldview of ocean and its inhabitants. This is not reclamation, nor is it rethinking the role of fish in histories of culture or economics or even a critique of how things have been done; rather, it is a new vision of recreational fishing and the fish that now inhabit the ocean, bays, and marine estuaries.

I adapt the title of this chapter from Stefan Helmreich's remarkable book *Alien Ocean: Anthropological Voyages in Microbial Seas*, in which he shows through an anthropological examination of marine microbial biologists that "life is strange, pushed to its conceptual limits, spilling across scales and substrates, becoming other, even alien to itself."[4] Recreational anglers must now ask just how alien fish are to us biologically, ecologically, and culturally, and how we might envision a new place for fish in the new ocean culture as we push the conceptual limits of how we understand fish.

Of course, taxonomically, each of us "knows" what a fish is, despite the ambiguity and difficulty of establishing a consistent, formal definition. The moment we encounter the representational word or image, we know what a fish is or what fish are. We each probably even maintain arbitrary, unacknowledged hierarchies of fish varieties: fish we prefer to eat, fish we find aesthetically pleasing, fish about which we are curious, fish that excite us, fish that disgust or frighten us, fish we want to catch, and fish we want to avoid. It is generally understood that "nature" is one of the hardest words in the English language to define, but for those who immerse themselves in Neptunian thinking, defining "fish" may be as difficult, despite an intimate familiarity with the object of such signification. In fact, we oversimplify our categorization of fish to the extent that thousands upon thousands of distinct species are reduced to mere "fish" with little concern for distinction. Some researchers and ethicists have indicated that we likely make such oversimplifications because of evolutionary similarities among species that inhabit aqueous environments: streamlined bodies, fins, and gills. Ruminate on that for a moment; imagine Christopher McKay's Alpha phoning you and asking, "What is a fish?" How might you construct your piscatorial definition for an audience with no reference point for analogy or comparison? Defining "fish," it seems, runs afoul of its reliance on contextual, subjective, and cultural influence. A fish and pornography—we know it when we see it.[5]

But sometimes we don't. Defining concepts like "nature," "pornography," "fish," or pretty much any other word comes down to a moment of power, a moment of naming a thing, and in naming that thing, establishing a place for that thing in the world. Think about this way: if we look at the simple, familiar situation of a mother rolling a ball to her young child, we can begin to understand that naming and defining are much more powerful acts than we tend to acknowledge. When the mother rolls the ball and says to her child, "Ball. Do you want to play with the ball?" she has exercised an act of power in naming the ball and the function of the ball (play) in the world for the child. She thereby creates a place for the object in the child's world, categorizing it and placing it in an ontological position. It is a thing of play, not to be taken too seriously. Of course, the ball could never be anything but what we decide to use it for. The ball, despite its intrinsic shape, is not an orange or a melon;

its value is assigned by our use of it, not its position—known or unknown—as part of a natural system. To name something "fish" is not only to identify it as an organism that dwells in aquatic space, but to assign it a specific cultural, historical, and economic position in the world—at least in the human world. We name fish; fish do not name us. To do so requires ethical consideration.[6]

This concept repeatedly comes into play in important and material ways. For example, in the seventeenth century, the bishop of Quebec requested that the Catholic Church categorize the North American beaver as a fish so that his congregation could eat beaver on Fridays during Lent. Similarly, the church has offered dispensation to allow capybaras to be categorized as fish and eaten on Fridays. Common lore says that the Vatican granted this dispensation when sixteenth-century missionaries requested it in the face of limited food supplies, and when conversion rates fell because indigenous populations were not pleased with meat abstinence during Lent. Missionaries argued that because the capybara spends so much time in the water and because it has (partially) webbed feet, it should be considered a fish. Similar dispensations have also been allowed for alligators as recently as 2010.

On the one hand, we can easily dismiss these kinds of categorical indulgences, chalking them up to a particular contextual need and dispelling them as exceptions to the rule rather than standard operating procedures. But we also have to acknowledge the complexity of making such allowances, and the effect that such moves have had not only on the populations of beavers, capybaras, and alligators, but also on the economies that surround the commodification of each of these animals within the allocation of the church. Note, too, the long-term effects of these three- to four-century-old traditions on local cultural practice; in Venezuela, for example, capybara feasts are considered part of the pre-Easter tradition during Lent and Holy Week. Venezuelan hunting laws now manage capybara harvests, and capybara farming has become common, contributing to the economic value of the capybara as a harvestable and farmable species.[7]

Certainly, few of us are going to point to a capybara or beaver and think "fish," though some may think "food" or even "game." We do, however, need to consider not just what it means that part of the world is willing to categorize these two mammals as fish—a categorization many of us may find to be truly alien—but what that might teach us about how we adhere to our own definitions and taxonomies and the effects such cataloging has on the fish themselves, on how we think of fish, and in turn, on the global ocean.

On May 9, 1980, the radar of the Japanese bulk freighter *MV Summit Venture* failed while negotiating the channel under the Sunshine Skyway Bridge at the mouth of Tampa Bay. The *Summit Venture* collided with the bridge, causing 1,400 feet of the steel cantilever bridge to collapse 165 feet into the bay, taking with it six vehicles and a Greyhound bus and killing thirty-five people. Thereafter, the bridge was unpassable. The new Sunshine Skyway Bridge was constructed between 1982 and 1987 adjacent to the old bridge site. At both ends, approaches to the old bridge were left in place. At first, these old bridge spans were unkempt, ignored by the city but coveted by anglers, who would drag old shopping carts of tackle, bait, and gear along the bridge in order to fish the channel that ran under the new bridge. In 1994 the Florida Department of Environmental Protection converted the old bridge into the world's longest fishing pier, given the reputation the unattended bridge had earned as a great location for catching many inshore and nearshore species: snook, tarpon,

Spanish mackerel, king mackerel, grouper, black sea bass, cobia, sheepshead, flounder, and red snapper. But before the pier became the pier and was still just a derelict road to the channel, anglers walking out onto the bridge were required to dodge not just potholes, trash, broken bottles, tires, and abandoned camp fires, but piles of fish carcasses strewn across the roadway and left to decay in the Florida sun. Before it officially became a pier and was still just a place from which to fish, I often walked out onto the remnants of the old bridge on the north side of the mouth of Tampa Bay, where the Gulf of Mexico flows in and out of the bay, stepping among the bodies of the dead, worthless fish, comparing them with the thirty-five human lives lost on that day.

If I had taken census of the dead fish, I would have logged assortments of gafftopsail catfish, skates, rays, sea robins, lizardfish, and endless numbers of greenbacks (also called pilchards) and other white bait (a generic term used to refer to many kinds of small baitfish), perhaps a handful of small jacks. I would have been less likely to find the bodies of mackerel, snook, grouper, cobia, and the like. Reductively, the bodies I would have encountered might all be cataloged as either "trash fish" or "bait," two terms that carry baggage of limited use-value, and therefore the willingness of anglers to simply toss these fish to the ground along with the broken glass, crumpled food wrappers, and dirty diapers that littered the bridge. These fish were waste, rubbish, the recreational fishing equivalent of what we identify in commercial or industrial fishing as "bycatch"—the animals we unintentionally catch and have no desire to make use of given their lack of assigned value. Since none of these discarded fish carried value as either game fish or food fish, they could be disposed of as thoughtlessly as one might toss out a candy wrapper. Perhaps more telling, too, is the ease with which anglers were able to make the ethical decision to toss the unvalued fish onto the bridge rather than back into the water, as the fishes' deaths were more convenient than their survival. The attribution of "trash fish" carries more deeply seated cultural relevance, as well. Many of the fish that the pier anglers might have categorized as trash were also prized by some minority populations as food, often even as delicacies. Because some minorities kept some of the "trash" fish, many anglers identified that as contributing to or confirming the worthlessness of the fish, their value bound up in a human racism reflected in the killing of fish valued by the devalued. ("Only a Filipino would eat that garbage," I've heard rationalized many times; or replace "Filipino" with derogatory terms for Mexicans, Cubans, Haitians, Chinese, Koreans, Japanese, and so on.) That is, it is impossible for us to begin to consider what constitutes fish, or even fish of value, without recognizing that race, class, and culture will always play a signifi-cant part in establishing such definitions. To clarify, a fish is not a fish just because it is a fish; it is a fish because it has under-gone human scrutiny bound by the baggage that such scrutiny always brings to the game. Often, such audits result in trash-fish mentality, the willingness not only to designate a species as being of little or no value, but to provide absolution for the wholesale slaughter of that species because it is simply trash to be disposed of, not a living organism, not recognized as part of larger marine ecosystems or somehow bound to the fish to which we ascribe higher value.

Debris. Not a discarded pallet floating on the water, but the bodies of living organisms. Trash. Rubbish. These are markers of status: what one can and cannot afford to discard, or at least what one perceives one can afford to discard. One person's trash is another person's treasure, they say; so, one person's trash fish is another person's food fish, but only when one can afford to throw out that fish. In regions where hunger prevails, the idea of trash fish is less pervasive because all fish are food fish—a value with its own set of ecological and sustainability difficulties. We can have trash fish only if we are well off enough to be selective. Being well off, though, does not necessarily mean financially. We may be well off simply because we can discard a fish knowing that the waters in which we fish are likely to produce a fish we value more highly, whether on culinary or sporting scales. Accepting one person's trash fish as your food fish grows from a condition of lack, or what we might more easily refer to as being poor, or simply having less. The starving person is less likely to discard the jack onto the pier deck unless he or she is confident that the water will provide a higher-value fish. That is, a good deal of our approach to understanding what constitutes "fish," or the range of values we attribute to fish, is consequentially economic. Rethinking the economic positioning of what is and is not a fish will inevitably be central to how we might imagine the future of the world's ocean. To be complacent in throwing a fish onto the pier to die is at once an ethical decision about the value of a specific life and a confidence in the provision of other resources. Both are pompous positions from which to operate. Such mind-sets will also be difficult to overcome, as they are tied to deep cultural and economic values and a number of perceived entitlements. Already I imagine the familiar critiques waiting to be levied against my position: "it's just a fish," the common dismissive that establishes a framework of value based not in ethic but in disregard; and "no one has the right to tell me what fish I can and cannot kill," the less-than-critical position born of arrogance and refusal to see ecological values.

And let's be honest here: the very concept of recreational fishing or sportfishing emanates from privilege. Recreation and sport—fishing or otherwise—can be taken up only by those who can afford to recreate. Those humans around the globe struggling for survival are not apt to take Saturdays off to dangle a line—and it would be despicable to suggest that they might be better off if they did. In our honesty, let's also admit that there is something rather unnerving (if not gauche) in having to say out loud that the $27 billion industry that surrounds saltwater fishing is an industry of expendable income, an industry of recreation and leisure, an industry that, yes, supports hundreds of thousands of jobs and family incomes. But really, it is a sign of excess, of our privilege as Americans to manage and value an activity that will always be beyond the economic reach of most of the world, and most likely outside of most of the world's epistemological or philosophical framework. We are astonished by the prince in Dubai who spends $18 million on a license plate; we can't grasp having the wealth that would allow us to throw away that much expendable income on a license plate. Imagine how much of the world would see our carefree willingness to buy a $100 fishing rod, a $10 lure, or even $5 worth of bait we will never ourselves eat. To fish recreationally is to embrace our privilege. We must willingly acknowledge this to ourselves each time we fish, expressing to ourselves and to the world our gratitude for such privilege. We are fortunate to live in a country where the very idea of a $27 billion recreational fishing industry can even be fathomed. This does not grant us entitlement to the resources bound to our fishing, to fish and ocean. Instead, in thinking about the new ocean, we must take up this place of privilege as a place of responsibility and leadership, nationally and individually, and adopt a rhetoric of responsibility. This means that each choice we make as anglers must be acknowledged as a choice of privilege, a choice that not all can make. The bodies of the trash fish we step around or tease each other about catching should be reminders that we subscribe to such values of trash and debris because we are privileged enough (and pompous enough) to pretend those values are inherent and not the result of privilege, value, and the willingness to imagine a fish as trash. And we must be open to considering, if not a quest for viable answers, then at least the willingness to accept the philosophical queries that surround this fishing life. What, for example, is the ramification of recreation morphed into industry? Am I willing to consider what it might mean to acknowledge that we subdue life for recreation? Also, what is the value of that life among trash and debris within our collective mind-set?

Fortunately, the Florida Department of Environmental Protection now manages the Skyway Pier to keep it clean of diapers and decaying bodies, but attitudes toward trash fish are not as easily sanitized. Of course, the Sunshine Skyway Pier is not unique in its history of trash fish. At various times I have seen the same disregard and killing on the Jacksonville Pier (before it was destroyed by Hurricane Floyd), the pier at Sebastian, Del Rey Pier, Pier 60 in Clearwater, Redington Long Pier, Cocoa Beach Pier, Dania Fishing Pier, Fort Myers Fishing Pier, Pompano Beach Fishing Pier, and the Sunglow Fishing Pier at Daytona Beach. I could easily expand this list to include piers I have fished in Virginia, North Carolina, South Carolina, and California, as well as those in states in which I have not fished from piers.

Piers are convenient locations to observe such willing discard, primarily because they are more readily accessible than many fishing spots, and they tend to draw a population of anglers who might not consider themselves as privileged as I suggest we all are, despite their ability to take some time to go fishing. The mentality of value assignment as fish or trash extends well beyond the pier to the very philosophical being of the recreational angler. These are the ethical deliberations we all make tacitly, rarely considering their implications on our larger allocations of value. These are the ethical positions we will all need to reconsider more deliberately as we reformulate the role of the recreational angler and recreational fishing in the next iteration of the world's ocean. This does not imply that we will need to think of previously otherwise-valued fish as food, or that the very idea of food or sport will even be at the center of the ethical valuation of fish. These are the ethical meditations that will contribute to the cultural value assigned to fish and ocean. When these circumspections emerge from within the culture of recreational fishing as counterpositions to the economic logic of the commercial and industrial harvest cultures—counter to food and sport as the primary, if not the only, worth of fish—we will be able to adapt more oceanic and sustainable thinking to our recreational fishing ethic. Encumbered in such thinking, we will inevitably find the difficult need to overcome centuries of assumed entitlement to take at will the lives of fish as a natural right.

A brief meditation: within the competitive roots of recreational and sport fishing, anglers have adapted a conventional competitive approach to the catching of trash fish. The aptly named "trash-can slam" provides anglers the opportunity to claim a kind of inglorious victory at catching a variety of "worthless" fish. The Grand Slam, a competitive signification of a successful safari in a single day of angling, is generally localized to reflect the angler's ability to seek out, hook, and land three or four of the region's gamest fish. The Florida Keys Inshore Grand Slam, for example, consists of a tarpon, a bonefish, and a permit all taken in a single twenty-four-hour period. Add a snook to the mix for the Super Grand Slam, the equivalent of the rare bases-loaded home run. The Offshore Grand Slam can be a marlin, sailfish, and tuna or other billfish, depending on the region.

The International Game Fish Association manages slam awards, which were introduced to sportfishing as a method for encouraging anglers to target multiple species in a given outing rather than focus on harvesting multiple specimens of a single species. According to the IGFA, Grand Slams are awarded for catching three approved species in a given day; a Super Grand Slam is four fish, and the Fantasy Slam comprises five caught fish. The IGFA identifies Billfish Grand Slams, Inshore Grand Slams, Offshore Grand Slams, Salmon Grand Slams, Tuna Grand Slams, Shark Grand Slams, and two freshwater Grand Slams: Trout and Bass. The IGFA does not, however, acknowledge the trash-fish slam, which is a localized version of the grand slams for catching trash fish: ladyfish, skates, catfish, some jacks, sea robins, toadfish, lizardfish, stingrays, pufferfish, and until recently, even barracuda. Yet, what happens when we create such categories and make a joke of identifying some species as trash, even providing awards at tournaments for trash-can slams as acknowledgment of an angler's success in failure? Is the trash-can slam a starting point for the revaluation of fish?[8]

It might seem that my position is angling toward an empowerment of fish, and in some ways that might be on point. However, I want to be clear that I am not arguing, as some have regarding other species or animals in general, for a human-equivalent subject position for fish. That is, I am not arguing for fish as citizen. I am not anthropomorphizing fish. What I am doing—and will continue to do—is arguing for a more dynamic way of understanding fish and the act of catching fish beyond just economic value. This is not an argument to be cataloged within an "animal rights" doctrine. It is an argument toward cultural repositioning of fish. It pushes toward a future of understanding fish and fishing in more complex ways than we have. It sees the very idea of "trash fish" as a demarcation with more encompassing cultural ramifications, as is the case with many other cultural-taxonomic labels we use.

To clarify, consider one species of fish not bound to the trash-fish label, and with little value to the angler beyond occasionally serving as bait. A fish with absolutely no human food value, yet a fish that H. Bruce Franklin has aptly identified as "the most important fish in the sea":[9] the menhaden—a fish that I would argue we might rethink in alien ways in order to develop strategies for rethinking fish in general.

As a boy, I would watch the surf and the water just beyond the breakers, learning to read them as one might any media. I checked for visual incongruities, nearly impalpable distinctions in the surface patterns and flow: nervous water, ripples, splashes, slashes, wakes, shadows, fins, tails, and birds, learning to discern the kiss of wind on the surface from a school of bait just beneath. I looked for the subtle signs that would expose the presence of fish below. Of all the signs and symbols the water would reveal, my heart would race most at the sight of a large dark mass moving through the water. Like the shadow of a cloud, only darker and more tightly formed, the nebulous shape would range from a few hundred feet across to thousands of feet. The water above the shadowed spot would offer slight glimpses of spectral reflections, a hint of oil on the water reflecting in the sun. If I were patient in my observation, it was likely that the water in and around the aphotic area would suddenly erupt in a frantic display of panic and relentless forage. The darkened water, a school of menhaden, packed together for defensive purposes, attacked and herded by bluefish, sharks, trout, jacks, redfish, and other game fish, alone or in schools. The attacks on occasion so merciless, the menhaden would flee through the surf onto the beach, flopping and dying on the sand rather than facing the predators demolishing their school. The episode lathering the surface white and oily.

In my youth—and perhaps to an extent even now—when I would witness the spectacular display of such feeding I would be wrapped in the oceanic sublime, the majesty tinted in terror, seeing the schools as vast sites of life and death, unable to truly grasp the expanse or relevance. The schools might be so big that I could barely comprehend what I was looking at, unable to imagine the sheer numbers of individual fish populating the collective. Other times, they would be broken into smaller, definitive clusters by the paths and boundaries carved out by the feeding fish. Attempting to calculate or imagine their numbers was conceptually amazing in ways akin to trying to comprehend size and numbers of star-filled sky. Uncountable flecks of light shimmering in spatial media. With age (not maturity), I learned to think apart from the sublimity, never unencumbered from it but adjacent to it, and to exact perspective. These schools, darkening the water's surface in what seemed to be immense shoals, were but pinpoints on the ocean's surface and but mere fractions of the menhaden population that ranges from the Yucatán to Nova Scotia. My own innumeracy still stymies my ability to perceive the vastness and quantities of menhaden that inhabit the ocean, let alone imagine the numbers that once ranged there.

If I saw a school unimpeded by attack, I would pull a large treble hook through the mass, snagging menhaden to use as bait. However, what I waited for primarily was that moment when the surface would rupture and in the midst of crashing white water, menhaden would rain down in all directions, fleeing in Darwinian hysteria from the teeth of predators besetting the school. The feeding, uncontained by the water, would shift media into the air above the surface as prey fled the attack below and the predator launched in chase. The commotion of these bodies raining upon the surface would sound as though a squall had suddenly unleashed a torrent of precipitation upon a confined surface. The sound loud with the tearing of surfaces and the strikes of bodies on the aquatic plane. The sound booming in the salt air, but upholding a kind of simultaneous silence, as the fish themselves made no sounds other than the slapping of their bodies. I've often wondered what it would be like if bluefish or jacks or other predatory fish roared and snarled in their attacks like bears or wildcats and if bait screamed and panted in terror. The shrill of birds taking advantage of the scene is the only animal sound unique to the moment. The silent animals feeding and fleeing, though, create the telltale sound of life and death, somewhere between cacophony and reticence. The sound and sight of the attack triggering my response in a cast, rod tip whipping the air, Hopkins spoon tracing an arc in the sky and splashing into the chaos of the feed. A steady retrieve with twitches of rod tip, recovering and dropping the lure through the bedlam back toward shore. And, on those treasured occasions, the reward of a sudden, voracious strike and the struggle to follow, a tug-of-war born somewhere in tension of sport and survival.

What I did not realize then was the global and ecological importance of what I regularly witnessed in schooling menhaden. These small, oily baitfish have come to represent for me more encompassing understandings of what constitutes "fish," and how the human relationship with menhaden stands as indicative of more capacious understandings of fish. Admittedly, Franklin's *The Most Important Fish in the Sea* opened my eyes to menhaden; it is one of those books I recommend as a must read, not just for anglers but for anyone interested in the future of the world's ocean. I suppose, too, that I find a kinship in Franklin's book given that he is, as I am, a professor of English, and I find in his work validation that English professors can have voice in the realm of ocean preservation and fisheries management.[10]

My reductive account of menhaden here does a disservice to Franklin's extensive research,[11] but I employ his account (and those of others) here as a starting point both to educate about menhaden and to appropriate them as a port of departure for addressing more encompassing attitudes toward and practices in managing fish qua fish.

While watching schools of menhaden up and down the beach during those boyhood days, had I been at the helm of a boat and pointed the bow just west of due north and traveled up the Chesapeake Bay a mere sixty-five miles, I would have found myself in a small inlet just north of Fleeton Point off of Ingram Bay and up Cockrell Creek south of Reedville, Virginia. Adjacent to the Reedville Airport is a series of docks owned and operated by Omega Protein, a century-old industrial fishing operation.

Omega Protein's entire operation is devoted to the harvesting and processing of menhaden to produce fish oil, fish meal, and proteins used as dietary supplements and ingredients in animal and human food. Founded in 1913 and now based out of Houston, Texas, Omega Protein harvests approximately 90 percent of the menhaden taken from US waters, a near-total monopoly. The harvest exceeds the tonnage of all other fish commercially harvested in the United States—combined.[12] Omega Protein is a publicly traded company (OME) that in 2014 earned $308.6 million. To provide a bit of perspective, note too that according to Franklin, since the 1860s, menhaden have been the nation's largest annual fish harvest in numbers of fish taken, second in tonnage only to Alaskan pollock. Larger than red snapper, haddock, cod, tuna, and salmon.

Omega's harvest techniques, as Franklin and others have described them, are frighteningly efficient, mechanical, and designed for maximal yield in minimal time. Its methods evoke science fiction–like resource stripping, as in the Martian fighting machines of H. G. Wells's *The War of the Worlds*, which snatch up humans, conferring them to a "great metallic carrier which projected behind him, much as a workman's basket hangs over his shoulder." Franklin and coauthor Tom Tavee describe Omega's harvest method in their *Discover* magazine feature article "The Most Important Fish in the Sea," which predates Franklin's book by six years:

> As the pilot approaches, he sees the school as a neatly defined silver-purple mass the size of a football field and perhaps 100 feet deep. He radios to a nearby 170-foot-long factory ship, whose crew maneuvers close enough to launch two 40-foot-long boats. The pilot directs the boats' crews as they deploy a purse seine, a gigantic net. Before long, the two boats have trapped the entire school. As the fish strike the net, they thrash frantically, making a wall of white froth that marks the net's circumference. The factory ship pulls alongside, pumps the fish into its refrigerated hold, and heads off to unload them at an Omega plant in Virginia.[13]

On the one hand, such technological efficiency has to be admired; on the other, we have to ask about the ramifications of such abilities for both the future of the world's ocean and, inextricably, for human culture (and, of course, the relationship between the two).

As Tavee and Franklin explain, menhaden are primary forage for legions of fish species. So much so that the nineteenth-century ichthyologist G. Brown Goode, who headed up the fish research program of the US Fish Commission and the Smithsonian Institution from 1873 to 1887, is quoted as famously proclaiming that "people who dine on Atlantic saltwater fish are eating 'nothing but menhaden.'"[14] As provocative as such a claim might be—and true to an extent—it only hints at the importance of menhaden in a larger framework of oceanic ecology. Menhaden, that is, should be understood not just as sustenance for the food fish humans eat, but as a critical biolink in the overall oceanic system.

Menhaden, members of the herring family, are filter feeders; they eat by straining suspended matter and particles from the water. As such, menhaden modify the waters in which they swim and feed, cleaning it of microorganisms and other tiny foodstuffs. Menhaden, as herbivores, feed primarily on plankton found in midwaters. They consume immense quantities of phytoplankton, contributing directly to the control of algal growth in coastal waters, sometimes helping to control the spread of red tide—algal blooms that often cause fish kills in affected areas, some qualifying as hypoxic.[15] As a filter feeder, each adult fish is capable of filtering four to eight gallons of water per minute, or nearly ten thousand gallons of water per day. Two menhaden, that is, could filter the entire volume of the average American backyard swimming pool in a day. Multiplied by millions of individual adults, the filtered volume of ocean water is enormous. Thus, menhaden's ecological value extends beyond being an intermediary energy exchange mechanism between primary producers and higher-order consumers; their modifications to surrounding physical environments affect both the environmental structure and the populations that inhabit those environments.

Nevertheless, Omega Protein's harvesting efficiency threatens to sever this biolink. In fact, in December 2012, the Atlantic States Marine Fisheries Commission (ASMFC) was, for the first time, forced to impose limits on Omega's harvest numbers; the commission capped "the total annual commercial catch at 170,800 metric tons, about 80 percent of the average harvest from the last three years."[16] Omega responded by threatening to cut approximately 10 percent of the jobs at its Reedville location when Virginia governor Bob McDonnell signed into law a bill that ensured compliance with the ASMFC regulation. For Reedville and the 250 Omega employees there, the cuts would be significant. Of course, the Omega threat was retaliatory and an attempt to create public opinion that Omega was the victim of overreaching government. Omega Protein, by the way, is the only commercial reduction fishery on the US Atlantic coast.

Since World War II, the industrial harvest of menhaden has continued to expand, providing menhaden for two markets: the reduction market and the bait market, the latter providing bait for both limited recreational fishing and the commercial Atlantic crab and lobster fishery.

Franklin's book provides a detailed account of the relationship between the menhaden fishery and the economic, political, and cultural histories of the United States. As an informative work that reveals much about a significant part of American (economic) history, Franklin's book offers detailed insight into an often-overlooked aspect of fisheries management and American history. Rather than summarize Franklin's book here, I want to take his lead and continue exploring the menhaden situation as an avenue to better understand the alien nature of how we think of fish and harvest. To do so, let me offer a few short claims.

First, the history of menhaden harvest shows how our entrenched mind-set about fish is bound up in tonnage-thinking. We understand and measure fish populations in weight rather than numbers, and we think about a population as a mass rather than a collection of individual organisms. This amalgamation of individuals into a singular conglomerate devalues the life of the individual. This is specifically why the death of the individual menhaden (or "trash fish" tossed onto the pier, if we want another example) can be tolerated—or more accurately, disregarded—because that "fish" is only a fragment of the imagined endless numbers that collectively make up what we understand to be "menhaden" (or "catfish," "ladyfish," and so on). Menhaden, in our cultural mind-set, refers not to an individual but to a collective. There is no individual menhaden. Menhaden plural is conceptually menhaden singular. The individual is lost to the cultural imaginary.

Such thinking is unique to our assessment of marine animals, and fish specifically. We do not count populations of terrestrial game in weight. Nowhere can I find estimates, for example, of how many pounds of venison roam wild in the United States or how many pounds of wild duck are harvested each year in the United States. We can perhaps attribute this difference in measurement to the fact that the commercial harvest and sale of fish well exceeds that of terrestrial wild game and that wild fish harvest grew to industrial proportions long ago while wild land-animal harvest did not, instead evolving domestication and farming practices to answer harvest needs. But such attribution also reveals much about our cultural valuation of fish compared with other wild game—as individual organisms (there are approximately 30 million deer in the United States) rather than collective sum totals (in 2015, the total allowable catch for Atlantic menhaden was set at 187,880 metric tons), for example. To highlight this distinction, Alison Mood, writing for the British organization fishcount.org in 2010, mentions the United Nations' Food and Agriculture Organization estimates that between 1997 and 2007, 92.2 million tons of fish were harvested per year, and that commercial fishing killed between 1 and 2.7 trillion individual fish each year. In her report "Worse Things Happen at Sea: The Welfare of Wild-Caught Fish," Mood argues that there is significant value in seeing fish as individuals rather than as a collective mass.[17] She notes, too, that this estimate is confined to commercial and industrial harvest and does not account for recreational fishing, or more important, illegal fishing, bycatch, or ghost fishing. Her objective in seeing fish as individuals, not as an assemblage, is not grounded in the unreasonable position to end fish harvest, but to develop an ethical approach to harvest.

Interestingly, in their proposal to develop standardized global recording and reporting of recreational fish harvest, Steven Cooke and Ian Cowx waver between using numbers of individual organisms and weight in reporting their findings. For example, they note that using their data, we can estimate that 47.1 billion fish are landed annually by recreational anglers worldwide, of which 36.3 percent are harvested, weighing about 10.86 million tons.[18] They offer these numbers to argue that overall global harvest needs to be increased by about 14 percent to account for recreational harvest on top of commercial and industrial harvest.[19] The difficulty here is not in Cooke and Cowx's use of numbers, but rather in the attempt to adapt those numbers to a consensus approach that accounts for both scientific needs and ethical understandings of the populations.

Ultimately, what I drive at with this first claim is a simple position: the recreational angler's respect for the life of an individual fish must drive the larger cultural perception of fish as individual organisms with intrinsic value, not merely as tons of meat. This claim asks us to apply an angler's ethic to an already problematic worldview that not only overlooks the value of the fish as an individual organism, but oversimplifies the very questions of how those values are established and culturally reinscribed.

Second, Omega Protein—like all commercial and industrial fishing operations—harvests and profits from a resource collectively owned by all citizens of the United States. These resources are best understood as the commons, property of the citizens of the United States. Other resources of the commons include some timber, minerals, and oil, and of course, other wild animal populations. However, unlike the timber and oil industries, Omega Protein and other commercial and industrial fishing operations profit from the extraction and sale of the resources of the commons without paying compensation to the collective ownership of those resources—the citizens of the United States. Oil, timber, and other industries pay stump fees, royalties, and leases in order to extract resources. If we are to preserve the world's ocean and its populations as part of the global commons—or even US waters as part of the American commons—then we can no longer allow this kind of privatization of common marine properties. And perhaps, to push this claim a bit further, we also need to adjust the very underpinnings that have allowed us to imagine ocean and fish as property to be claimed, distributed, and owned. Instead, the angler's ethic ought to recast the ocean as a global region critical to all humans and stewarded by all humans.

Third, though the fishery itself is monopolized by Omega Protein, regulations that govern menhaden harvesting are established, monitored, and enforced by governmental agencies—at the state, regional, national, and even international levels. These agencies depend on information from multiple sources to inform policy decisions, including information from the industry itself, from independent scientific sources, from within the governmental agencies themselves, and often from the community, including recreational and professional anglers. Thus, the very notion of fisheries management and conservation is predicated on the politics of communication between multiple invested parties, often with differing data, motivations, agendas, and beliefs—even as to what a fish is (commodity, organism, resource, etc.). These relationships must be rethought, reconfigured to privilege scientific data above economic impact as the primary influence in policy configuration. This is particularly true when we place the privatization of the menhaden fishery alongside similar privatization maneuvers such as the development of the catch-share policies that have evolved over the last decade, as I will show later.

Fourth, by identifying that the menhaden fishery is framed, discussed, and conceptualized, as well as regulated and monitored as a "United States fishery," we must also acknowledge that our thinking about fishery management is predicated on a land-based notion of political boundaries, which may be less useful than thinking of fish populations in terms of ecological boundaries. Fish, that is, probably do not care or even comprehend whether they are in US waters or Canadian waters. I am not arguing here for the kinds of environmental mapping proposed by bioregionalist thinkers such as Allen van Newkirk, Peter Berg, or Raymond Dasmen, or even the ecoregions defined by the Environmental Protection Agency or the National Forest Service, which are grounded in stagnant, fixed land-based thinking and are inextricably bound to a foundational notion of national and political borders. Instead, we need to imagine a more fluid concept of fluctuating ranges of tolerance, tied not only to regions but to fluid shifts in oceanic contexts such as temperature change, current flow, salinity levels, currents, and migration routes. The very kinds of alien thinking that leave us bewildered when tropical fish such as permit turn up in Louisiana waters, or when Southern California and Mexican tuna species become familiar to Oregon waters, lead us to conclude that these movements are anomalous rather than intrinsic to a fluid environment that does not obey the rules of land-based boundary logic and politics.

Fifth, and tied directly to the previous claim, familiar ways of thinking about fish and fisheries tend to localize and limit our view of fish. Fisheries management, for example, generally monitors and regulates local populations; management policies are developed based on local observations and are purported to benefit local fisheries. On the one hand, localized thinking is critical in understanding local populations and moving toward local management; however, localized thinking can also blind management thinking toward larger ecological and global concerns. Strictly localized thinking is not conducive to monitoring or regulating global fisheries issues. Localized thinking risks ignoring larger ecological factors influencing fisheries. For example, red snapper allocations in a single Gulf Coast region should take into account not only red snapper harvest allocations throughout the red snapper range of tolerance, but also other practices and policies that affect connected ecosystems, such as estuarine shrimping, which contributes to forage availability for juvenile fish, including red snapper. Fish, that is, are not exclusively local, even if the fish appear local to those who harvest them.

One example of limited localized thinking emerges in what we might call closure mentality, the imposition of policies that deny recreational anglers access to specific local waters. In areas heralded under the banner "marine protected areas" (MPAs), management agencies often close local fisheries under the assumption that closures will allow a region to recoup. The difficulty here is twofold: first, closing one area does not eliminate anglers from their activity. Instead, anglers find other areas to fish, thereby increasing pressure on other local regions. Second, closing one specific area assumes that region to be autonomous in its ecological relationships with neighboring regions and the global ocean in total. Reefs—or other defined locations—are not capable of isolationism; their boundaries are enigmatic, shifting.

Likewise, many MPAs are imposed—as was the Biscayne National Park MPA in 2015—based not on scientific evidence that closures have any sort of rehabilitation effect on a region, but on the belief that logically, closures *should* afford reclamation. That is, closures based on desire to protect local fisheries are, in fact, closures based on philosophy and ideology rather than on evidence. Yes, it may seem that closing one area to fishing should logically provide opportunity for improvement, but that is the case only if that region is a closed system not tied to vastly complex ecosystems, which no marine environments are. From redistributed fishing pressures to effects on migration and reproduction patterns, any local region is always already connected to more elaborate, broader-ranging ecosystems. Thus, limited local thinking in terms of fisheries management can be shortsighted in multiple ways.[20] Consider, for example, the attempts to manage menhaden harvest locally in Chesapeake Bay.

In January 2017, Virginia house delegate Barry D. Knight (R) introduced "House Bill 1577 Management of Menhaden," which would require the

> Virginia Marine Resources Commission to adopt regulations to implement the Atlantic States Marine Fisheries Commission Fishery Management Plan for Atlantic Menhaden and authorizes the Commission to adopt regulations for managing the Commonwealth's menhaden fishery. The bill also requires that any moratorium on the fishery be subject to legislative review. The bill repeals several Code sections relating to quotas, allocation of allowable landings, and administrative procedures that will be included in a regulatory framework for managing the fishery.[21]

But the bill was sent to the Chesapeake division of the House Agriculture, Chesapeake and Natural Resources Subcommittee, where it has remained without any action. However, in February 2018, Knight introduced a more specific bill identified as "House Bill 1610 Menhaden, Total Landings," which would adjust

> the annual total allowable landings for menhaden upward from 168,937.75 metric tons to 170,797.17 metric tons and provides that any portion of the coast-wide total allowable catch that is relinquished by a state that is a member of the Atlantic States Marine Fisheries Commission shall be redistributed to Virginia and other states according to the Commission's allocation guidelines. The bill adjusts the annual harvest cap for the purse seine fishery for Atlantic menhaden in the Chesapeake Bay downward from 87,216 metric tons to 51,000 metric tons. The bill also removes a provision that applied the amount by which certain actual Chesapeake Bay harvests fell below the harvest cap as a credit to the following year.

The bill, which Democratic governor Ralph Northam supported despite Omega Protein's resistance, would cut the overall menhaden harvest in Chesapeake Bay by reducing the quota by 41.5 percent.

However, in early March 2018, the debate over menhaden harvest in Chesapeake Bay took an unusual turn when the Virginia government and Omega Protein agreed to work together to convince the Marine Fisheries Council to develop a new quota that would be acceptable for both the local government and the industry. This kind of collaborative approach has not been common in fisheries management debates, but it holds potential for how management might move forward.

In 2017, the Mid-Atlantic Fishery Management Council

reduced the menhaden harvest quota in the bay by more than 36,000 metric tons, drawing Omega's ire. Previously, the standing law had identified the council's quota of 87,216 tons as the legal limit, but Knight's bill sought to increase this amount in coordination with the council's newer stock assessments and worked to remove the specific amount of 87,216 tons from the law. According to Knight and Matt Strickler, Virginia's secretary of natural resources, identifying exact figures such as 87,216 tons makes it easier to develop sanctions, including harvest bans.

Recreational sport anglers have also been vocal about the debate over menhaden harvest in the bay, siding with environmental groups like the Nature Conservancy and League of Conservation Voters to argue for lowering the quotas even further to protect the region. Recognizing the effect of harvest on other species such as striped bass and osprey, anglers have argued for stricter limits. Likewise, anglers worry that reductions in local menhaden populations may also affect the populations of migratory fish that feed in the area. Anglers point to the noticeable reduction in juvenile menhaden populations in the bay as indicative of larger regional and global problems tied to menhaden overharvest.

Yet when the government of Virginia and Omega Protein agreed to work together to propose feasible quotas to the council, Knight pulled H.B. 1610 from the House and the bill was never called to vote, much to the disappointment of anglers and conservationists. Knight argued that the collaboration between the Virginia government and Omega to persuade the council would be a welcome approach to negotiating fisheries debates.

Of course, Omega Protein touted the failure of the bill and the proposed collaboration as a victory, repeating its rhetoric that the bay was obviously healthy and that increasing harvest quotas would not affect its health status. Nonetheless, the collaboration provides an interesting approach to fisheries management. The problem, of course, is that the collaboration excludes a recreational angler or conservationist voice at the table. It also exemplifies the ways in which localized management thinking limits the understanding of a local problem, allowing it to be perceived as independent of global ecosystems and conservation efforts.

4

Ethics

Philosophy in the popular meaning of the word
has always been, and always will be,
central to recreational fishing.
– Robert Arlinghaus and Alexander Schwab,
"Five Ethical Challenges to Recreational Fishing:
What They Are and What They Mean"

One summer toward the end of the 1970s, when I was about
ten or twelve, my family spent a week in a cabin at Douthat
State Park in the Allegheny Mountains of Virginia. It was an
oven-like week of cutoff shorts and KISS T-shirts, soundtracked
with "A Rock 'n' Roll Fantasy" by the Kinks, Van Halen's first
album, Steely Dan's "Peg," and Gerry Rafferty's "Baker Street."
I took relief in the occasional cold orange soda—allowed only
because it was vacation time—and hours of splashing in the
park's sanctioned swimming area, where a jukebox jacked into
a treble-high PA system whined out the hits of the summer.
To this day, I cannot listen to the Atlanta Rhythm Section's
"Imaginary Lover" without craving a cold Nehi[1] and a bag of
Fritos—probably not the effect Dean Daughtry and Buddy
Buie[2] were looking for.

The lake at Douthat was and is stocked with rainbow,
brown, and brook trout and has strong populations of pickerel,
crappie, catfish, and largemouth bass. As an ocean brat and

germinating blue-water bum, I knew almost nothing about lake fishing (still don't), but I knew that I'd rather be fishing than not. I spent most of that week sitting on the bank by the lakeside cabin casting things I imagined to be appropriate lake bait: worms, crickets, bread balls, corn, and even a few of Uncle Josh's pork rinds that I had begged my mom to buy for me at the shop adjacent to the park.[3] I caught my share of crappie and little catfish that week, or at least enough to keep me interested and casting each day. Other anglers would walk past me, and some would stop and offer pleasantries and ask how the fishing was. Some carried spinning gear, some carried fly gear, and all talked about the trout for which the lake was famous. I, of course, wanted to catch one of those trout but stood little chance of doing so from my spot on the bank. The possibility, though, drove my youthful persistence.

On one occasion, a group of older boys, probably young teenagers, took up position up the bank from me and began casting baits into the lake. This was a boisterous group, smoking cigarettes, pushing one another around playfully, and generally participating in a teenage hullabaloo. Because I was younger, I was a bit intimidated, wanting them not to notice me, but not wanting to surrender my spot on the bank, either. Who knew what humiliation and torture a gaggle of teenagers would set upon me? I sat quietly, fishing and occasionally stealing glances at the rabble-rousers.

As a result of no skill of their own that I could observe, the hellions managed to hook and land a crappie. They hooted and hollered about their fishing prowess and tossed the fish back and forth like a hot potato. One of the boys—now the chieftain in my imaginative memory—grabbed hold of the fish, lit a match, and held it to the fish's eye. He ordered the others to be quiet and listen as he laid the fish in the lake and the burnt eye made a distinct sizzle as it touched the water. The boys erupted in violent laughter. They pushed each other and fell about the bank possessed by uncontrolled, mob-addled laughter. Somewhere between terrified and horrified, I crouched into the bank praying that they would leave, and even more so that they would not acknowledge me there.

I have held on to that memory for all these years not for the self-evident difficulties of the machismo violence and mob mentality of the moment, but because the act seemed to indicate not merely youthful ignorance and fierceness, but deeper cultural ideologies about the treatment of nonhuman animals, fish in particular. Perhaps in some inspirational manner, that moment provoked in me the need to be responsible for my own fishing in ways that honored rather than tortured my quarry.

When I teach my students to think about the hierarchies of value we assign animals, I ask them to participate in a simple mental exercise: S. C. Johnson and Son introduced Raid insecticide in 1956—six years before the publication of Rachel Carson's *Silent Spring*. Raid is currently sold in the United States, United Kingdom, Australia, and Saudi Arabia. It is identified as an effective pesticide against more than twenty kinds of insects. That is to say, it is an effective technology in killing particular types of insects. The brand's success over the last sixty-plus years also attests to consumer confidence in the product's usefulness. Raid's marketing slogan is "Kills Bugs Dead." The slogan was created by the advertising agency Foote, Cone and Belding—now one of the largest global advertising agency networks. The slogan is credited to Lew Welch, the American beat poet, who worked for the agency at the time. With these details of technology, writer, and paid slogan consultant in mind, I ask students to imagine a product successfully akin to Raid whose slogan is "Kills Chimpanzees Dead" or "Kills Penguins Dead." To this, I would now add "Kills Snapper Dead" or "Kills Menhaden Dead."

I designed the exercise to help students think about what kinds of animals it is culturally okay to kill dead and what animals it is not okay to kill. Of course, the exercise is deeply problematic because a simple advertising phrase cannot account for the entire cultural and historical nuance mixed up in determining what is and is not okay to kill. For example, we might not be comfortable with "Kills Cows Dead," but we are fine with "Billions and billions served." The exercise asks that we consider how reductive commonplace thinking can become when addressing complex ethical issues pertaining to choices we make about harvest.

Let me be clear again: my objectives in asking students to consider these kinds of cultural values are not grounded in an animal rights position; my stances do not argue against the harvest of animals. Instead, I encourage students and readers to be alert to the ethics of harvest and to become hypersensitive to the ethical choices and the histories that inform those choices. Such choices are never innocent and therefore require our individual consideration to such a degree that we can each justify to ourselves the reasons for making the harvest choices we make.

Hawaiian poet Joe Balaz, who writes in several languages, including Pidgin, published the thought-provoking poem "Gottah Eat 'Um" in *Hawaii Review* and then *Electric Laulau*.[4] It is a poem that I return to when thinking about killing fish; it is a poem I would like to have read against the visual backdrop of the Skyway Pier and the bodies of the discarded trash fish. It is a poem that I imagine hearing as Omega Protein strips menhaden from the water.

Wen I wuz wun small kid
My fadah told me
Anyting you kill, you gada eat
You shoot da dove
wit a BB gun—
you gada eat 'um
You speah da small Manini
at da beach—
you gada eat 'um
You help your madah
kill da chicken in da backyard—
you gada eat 'um
Whoa brah—
tinking back to small kid time
and da small kid games I used to play
aftah I heard dat
no moa I kill flies wit wun rubberband.

Balaz captures the spirit of ethical consideration that drives at the heart of the recreational anglers' harvest ethic: to kill or not to kill, and if to kill, then to what end?[5] At this point in our collective anglers' history, we should be able to say that the recreational angling ethic adheres to an explicit understanding that the only fish to be harvested are those to be eaten, those to be used as bait to catch other fish,[6] or those extracted for scientific research. In addition, we adhere to the strictest care and empathy in harvest methods. In fact, I propose this imperative to be the foundation of the new recreational angling ethical position. There is no need to kill for any other reason; to do so is not an inherent human right. This includes, by extension, the unnecessary slaughter of recreational bycatch, whether bait or trash fish. None of us can assume we are well off enough to discard any living marine organism, because we can no longer assume an infinite wealth of oceanic resource to provide better than what we dispose, nor can we operate as individuals making claims to such wealth. This kind of ethic, bound to catch-and-release philosophies, is by no stretch new to the angling mind-set. A. A. Luce's classic 1959 book *Thinking and Fishing* introduced the idea that angling might be understood as cruel if the intent of the practice was anything other than harvest for food. Luce, a professor of philosophy at Trinity College Dublin, began asking the very questions of ethics and angling that we must now revisit in order to prevail on anglers to consider a new ethic for a different historical and ecological context.

Perhaps more to the ethical point, too, none of us should have the hubris to assume that any one of us is entitled to kill whatever any one of us chooses. It is not *your* right to harvest; it is our collective privilege to access *our* resources (and there is difficulty even in regarding organisms such as fish as a resource rather than as individual living animals). The fish you kill is not your fish, and it is not my fish; it is our fish (or again, perhaps the very idea of possession here is misconstrued and the fish is, in fact, only the fish's fish). Assumed possession by the act of capture cannot be the driving philosophy behind an assumed right to kill. The recreational angler must now acknowledge a collectivity to harvest ethics. Hence, the very idea of what it means to fish recreationally should now imbue an always already catch-and-release mentality. Recreational fishing and sportfishing should be understood to be synonymous with catch and release. The right to decide to keep fish for nourishment should never be eliminated (or even questioned) as a recreational angler's right, but angler thinking should frame such decisions within a new culture that sees harvest as the minority objective, not the primary objective.

This ethical precedent requires much more substantial scientific and philosophical consideration of catch-and-release fishing. Unfortunately, legitimate considerations of the ethics of catch and release are too often tempered by many who misunderstand the role of catch and release and who lobby against it based strictly on ideology rather than on legitimate scientific evidence. And, to be fair, just as often, supporters of recreational angling rely on moralistic assertion rather than thoughtful engagement in order to demand "rights" to fish. These kinds of emotionally driven ideological positions do little to advance any legitimate engagement of the issues that surround recreational fishing. For example, Alexander Schwab's 2003 book *Hook, Line, and Thinker: Angling and Ethics* is such an intellectually flat, moralistic diatribe against animal rights positions as an avenue to support anglers' rights, that the argument Schwab proposes not only falls flat but paints the angling community as little more than a fundamentalist cult—a portrayal that risks more damage to anglers' rights than support. It is precisely the spread of such ignorant positions that does disservice to the community of anglers.

The narrative that unfolds in the first quarter of the twenty-first century concedes the peril of the ocean's future, rightly attributing primary cause to human intervention. Marine ecosystems of the Anthropocene are degrading at rates that outpace recovery efforts. Legitimate claims of hysteresis now pepper the oceanic narrative. For a decade now, that narrative has included concrete predictions as to when the saltwater fisheries will collapse.[7] The collapse narrative arises from long-term scientific research. Concurrently, the last decade has also produced an increasingly popular account regarding the agency of animal others, including marine organisms, particularly fish. This narrative grows from increased concern about treatment of nonhuman animals and from increased knowledge provided by improved technologies that furnish greater research access to the world's ocean. More-nuanced aspects of the conversation account not only for physical treatment but also for epistemological and philosophical treatment. That is, there is concern not only about animal treatment, but also about how we think about and place animals culturally. More-committed participants in this conversation argue for animal rights as being akin to human rights, thereby arguing for understanding animal subjects as equivalent with human subjects. This argument can be reductively characterized as taking two forms: the first emerging from academic and scientific research, and the second from a more popularized environmental and philosophical version championed by animal activist groups such as People for the Ethical Treatment of Animals (PETA).

If we are to embrace a recreational fishing ethic based on research and knowledge rather than on belief, then we must first acknowledge the value of the research that informs such an ethic. In 2006, Davis Guggenheim released the Academy Award–winning documentary *An Inconvenient Truth*, which documented the work of former vice president Al Gore to educate citizens about global warming.[8] Of course, "global warming" and "climate change" have become two of the most readily dismissed scientific issues in human history, despite the overwhelming scientific evidence demonstrating human impact on shifting global climates and the impending devastation such shifts will have on the planet. Guggenheim's title resonates with a cultural denial of the scientific claims regarding climate change. Because the truth—what I will more accurately refer to as the science—conveys information that is inconvenient to particular cultural mind-sets, that science is dismissed. Inconvenient science. The recreational anglers' ethic cannot dismiss inconvenient science. Even the science that offers conclusive evidence contrary to our assumptions, desires, or needs must be granted authority. However, this does not mean that all science is inherently good science or accurate science. Like all knowledge making, scientific discovery requires contesting, questioning, retesting, and refuting—but only through legitimate scientific inquiry. Consider, for example, my earlier identification of the difficulty of creating marine protected areas and closing those areas to recreational angling. Closing areas based on the *belief* that closures *should* aid reclamation is different from providing long-term evidence showing the need for and effectiveness of closures. Such evidence must inform the recreational anglers' ethic and acknowledge that the science displays a closure need in order to protect not just a local fishing region, but the very right and access to participate in recreational fishing. The moment we condemn scientific research as inconvenient or the moment we assume our individual right to harvest trumps the scientific evidence, we have surrendered our ethical position. This does not mean that recreational anglers must accept all science as accurate wholesale or that we must surrender our right to challenge the accuracy of science. However, it does mean that we respect the science even when we disagree with it or with how it is used in policy implementation. Only with scientific backing will recreational anglers be able to develop an ethic grounded in an ecological understanding of the relationship between fish, ocean, and angler; an ocean-centered cultural mind-set; and the cultural tradition of recreational angling.

Likewise, recreational anglers need to adopt critical interpretive strategies to consider how scientific research is used both to support antirecreational fishing policies and, perhaps more important, to feed antirecreational angling ideologies and cultural perspectives. For example, ethologist Jonathan Balcombe's *What a Fish Knows: The Inner Lives of Our Underwater Cousins* (published in 2016 by *Scientific American*), relies heavily on summarized and synthesized scientific data in order to overturn many long-held assumptions about fish. Balcombe's research is thorough—though often secondhand—and his argument sound. He shows readers a complex view of fish behaviors not considered before. His argument—to be overreductive about it—is that we have misunderstood fish as nonthinking, nonsentient creatures, but detailed observations of their behaviors reveal that fish can be and should be thought of as individual organisms with complex social lives, dynamic communicative abilities, and other sentient characteristics. Balcombe's research is insightful, and for those of us who wish to better understand fish qua fish, his book is quite educational. However, Balcombe, like many others, relies on an interpretive leap of faith that is deeply problematic.

Balcombe cites multiple research projects that clearly show that fish feel pain. That fish can feel pain or absolutely do feel pain has been central to arguments against recreational fishing, including catch-and-release angling. PETA, for example, is adamant in its publicity materials that catch-and-release fishing should be stopped because fish feel pain. It cites research from the University of Glasgow and the University of Edinburgh indicating that pain receptors in fish are strikingly similar to those in mammals and that "fish have the capacity for pain perception and suffering."[9] The claim that fish have the capacity for pain perception is a much different statement than PETA's interpretation that fish definitely feel pain. Of course, that is an evident problem. Balcombe more ethically represents similar research, identifying the accuracy and problems of saying a fish has the *capacity* for pain. But let's open the door to a more dynamic consideration and accept the claim that fish feel pain, but let's also consider what that might mean. When PETA claims that fish feel pain, the rhetoric implies that pain is synonymous with emotional trauma. PETA implies that pain is more than a physiological reaction to nerve stimulation; that is, it equates pain with human response to pain. It anthropomorphizes pain, and thereby fish. This, then, is a fallacious interpretation that relies on human arrogance to associate the physiological function of pain with a human emotion. Such false connections might seem logical, but they are grounded in human ego, which allows us to assume that all organisms function and respond as we might. Balcombe's approach is more thoughtful, more scientific, but nonetheless trapped by the hubris of anthropomorphization.

Similarly, Victoria Braithwaite's *Do Fish Feel Pain?* makes a cogent argument about fish feeling pain, thereby suggesting that humans ought to extend the same degree of empathy and care to fish as we do to other animals and birds. However, Braithwaite clearly understands the need for further scientific observation before reaching definitive conclusions about a fish's sentience. She explains:

> But while I agree that an aspiration for fish welfare is warranted by the evidence, we must be careful how we approach it and in particular we should be mindful that our understanding and the science underpinning what fish need or prefer is just a beginning.[10]

Braithwaite cautions against rushing to decisions about what we know about fish before we have all the evidence, which we do not yet have. I take Braithwaite's caution not just at face value to suggest that we need to know more before reaching conclusive ethical positions; it suggests that we also have to be willing to change our minds. That is, just as the ocean is fluid and shifting, we too have to be willing to shift our anglers' cultural baseline about fish as organisms and how we position them culturally.

To put it another way, Balcombe, Braithwaite, and other researchers are correct: fish feel pain. Accepting this knowledge is important in how we view the fish we pursue. Scientific consensus also demonstrates that it is highly likely that fish are conscious beings. We should accept this as well, and in doing so, allow such evidence to inform the recreational anglers' ethical imperative. In doing so, too, we should not find easy ways to dismiss such scientific evidence simply because it might force us to alter our perceptions of the fish we encounter. In fact, it is precisely because such research provides knowledge and insight that we should insist that it informs our ethical decisions. We cannot dismiss such findings out of hand simply because they are inconvenient within our angling paradigm. We should not ignore such research and offhandedly dismiss it as conveniently inconclusive or simply speculative, and therefore irrelevant or inaccurate. Instead, we must learn from such research to better understand our relationships with the fish we pursue.

Yes, fish feel pain, but how they respond to such pain is yet to be determined. It would be irresponsible to dismiss this claim by calling it irrelevant or simply by arguing that we cannot interpret fish pain as equivalent to or even related to pain as we understand it as humans. Such underthought dismissals serve no purpose. Rather, we need to consider the idea of fish pain within the context of fish lives to the best of our ability. Largely, we can never grasp the authenticity of the reaction a fish has to pain or to any other physical or emotional stimulus. Keep in mind, too, that while Balcombe and Braithwaite show us the legitimacy in claiming that fish feel pain, and PETA asks us to respond

emotionally to the *idea* that fish feel pain, we need to acknowledge that the physiological response to pain is not reserved
only for a wound from an angler's hook. In fact, fish feel pain
every day, and if they feel fear, as some argue, then it is likely
that they feel fear every day, as well. That is, pain and fear
are always already present fish responses to daily life. It is our
responsibility as anglers not to contribute to that pain or fear in
any significant way. We must develop a catch-and-release ethic
that works with and supports the scientific evidence we have in
hand and be willing to adjust to new information as it enters
the conversation.

We must also acknowledge that any recreational anglers'
ethic we develop must be derived within contexts that take
into account other pressures placed on fisheries and oceans. For
example, the recreational anglers' ethic cannot emerge without recognizing the impact of commercial harvest. In fact, it is
specifically within the relationship to commercial or industrial
harvest that we are able to identify recreational anglers' point of
view. Think about it this way: for a long time, the only access to
saltwater fish came through early forms of commercial harvest.
Fishing was conducted to provide protein (fish meat) to a population. Culturally, then, we established the very act of taking a
fish from the ocean as being tied solely to an act of consumption.
That is, the only reason one might fish would be to gather fish to
eat. Concepts of recreational fishing or sportfishing did not enter
into the cultural mind-set until very recently and emerged only
in relation to harvest for food potential. Until recently, the very
idea of catch and release, or more generally recreational fishing
writ large, has been incommensurate with cultural thinking.
Why would one fish if one did not keep and eat the fish caught?
This would be more than illogical; it would verge on unethical.

For a long time, the bottom line for fishing was food. It was not until June 1, 1898, when Dr. Charles Frederick Holder landed a 183-pound bluefin tuna on rod and reel out of Avalon on Santa Catalina Island, California, that the idea of fishing for sport began to take hold in the collective mind-set. But unlike river fishing or lake fishing, saltwater fishing held a mystique, driven primarily by technological restrictions on accessing saltwater fish and a long-held mythology about the dangers of the ocean. Following Dr. Holder's success, the world's first sportfishing club, the Tuna Club of Avalon, opened on Santa Catalina Island, California. The club not only provided cama-raderie among the explorers of this new sport but also worked to establish consensus among anglers as to how to take game fish in a fair manner. One of the primary objectives of the Tuna Club of Avalon and the clubs that followed was to establish codes of conduct and codes of ethics for saltwater anglers. As an early member of the club, renowned writer Zane Grey was not just central in circulating the heroic tales of saltwater angling that helped popularize the sport; he also helped estab-lished the fundamental values sport anglers were adopting.[11] Likewise, Ernest Hemingway was actively writing about big game fishing on the East Coast; his writing, like Grey's, worked not only to popularize the sport but to establish a code of ethics. Hemingway, like Grey, was instrumental in the begin-nings of East Coast fishing clubs. In 1936, he, along with other anglers, formed the Bahamas Marlin and Tuna Club. It was at this club that Hemingway penned the first dedicated rules of saltwater sportfishing:

1. All fish must be hooked, fought and brought to gaff by the angler, unaided.
2. Aid will refer not only to the angler but to any part of the boat or chair except the rod socket and harness.

3. The fish will be considered brought to gaff when the leader can be reached by the boatman.

4. The boatman may touch no part of the tackle except the leader, but he may adjust or replace the harness during the fight.

5. When brought to gaff a marlin or tuna may be subdued only with a club.

6. A mako shark is considered a game fish and should be gaffed or tail-roped.

7. The gaff may not have more than twenty feet of rope attached.

8. The rod or reel may not be fastened to the chair during the fight.

9. A reel falling off the rod butt, as well as a broken rod, will disqualify the fish.

10. No mutilated fish will qualify or be considered a legitimate catch.[12]

What is telling about these early rules of angling ethics is that they focus solely on the actions of the angler and the technologies the angler may employ. The fish itself retains no active role in the interaction, rendered an object upon which the angler acts. Of course, at the time, this was the understanding of the fish's part in the interchange, and the very idea of ethics in relation to animal others was culturally ensconced in questions regarding the legitimacy of using animals. Reductively, then, the traditional anglers' ethic might be considered an ethic of resource use.

More recently, anglers, philosophers, writers, and scholars have emphasized rethinking angler ethics—particularly catch-and-release ethics—in light of more contemporary fisheries and ecological contexts. J. Claude Evans, in his book *With Respect for Nature: Living as Part of the Natural World*, identifies that inevitably we eat at the expense of animals and that catch-and-release ethics can help us better understand our position in the natural world. Evans writes of catch and release that "we need more than ever to cultivate experiences and practices that connect us with integrity and wholeness. . . . The practice of catch-and-release is most properly based on respect for the integrity of ecosystems and populations that are subjected to the pressures of human use and exploitation. Embedded in this practice is a specific respect for the individual fish one attempts to catch and then release."[13] It was not until very recently in the history of recreational angling that any ethics developed regarding a more encompassing view of the fish, both as an individual organism deserving of respect and as a part of a complex ecosystem in which it lives and functions.

Evans outlines the critical aspect of a contemporary anglers' ethic, clarifying the importance of catch and release not just for the fish but for larger ecosystems, as well. Such recognition attests that catch-and-release angling affects both the individual fish and a more complex series of relationships with larger ecosystems—both immediate and distant. Ultimately, then, the contemporary anglers' ethic requires accounting for three primary facets of recreational fishing: (1) the actions of the angler; (2) the question of fish and animal suffering; and (3) the impact of recreational fishing on local and global ecosystems.

In one of the most rigorous considerations of recreational fishing ethics to date, Robert Arlinghaus and Alexander Schwab write that any ethical discussion regarding fish as individual organisms inevitably reduces the question of the angler-fish or human-animal relationship to a question of animal *use*. They address animal use from three key ethical perspectives:

animal welfare, animal liberation, and animal rights.[14] For Arlinghaus and Schwab, animal welfare refers to "taking due account of the potential impacts of our actions on the well-being, health, and fitness of individual fish and act[ing] in a way that minimizes these impairments."[15] They are also clear that there needs to be a distinction "between science-based, objectively measureable definitions of animal welfare from the issue of suffering in the world of fish and from ethics of animal welfare."[16] Taken in total, though, ethics grounded in animal welfare provide points of departure for considering how anglers might care for fish and reduce harm to fish when fishing.

Buttressed alongside animal welfare, Arlinghaus and Schwab point to animal liberation, by way of Peter Singer, as a kind of utilitarian philosophy grounded in questions about the results of human action. Animal liberation addresses, in part, the question introduced earlier regarding whether fish feel pain as the result of human action. For the animal liberationist, humans should not engage in any activity that might result in pain to an animal. Given the question of fish pain and fish suffering, and the scientific question of a fish's potential to feel pain versus the anthropomorphic sensation of suffering, animal liberation is complex and problematic. As I have said, as anglers we must concede that fish feel pain—or at least that they have the potential to feel pain—but we must also not equate that pain with human emotional responses to pain or to the ability to suffer physically and emotionally in the same way humans do. Thus, an anglers' ethic accounts for the animal liberation position through a definitive scientific approach, separating the anthropocentric view of fish and pain from the scientific evidence that shows the potential for pain—fish pain, not anthropomorphized suffering.

Turning again to Singer, Arlinghaus and Schwab explain that animal rights positions see all living things as having inherent value as individual living organisms. It is the animal rights position that most directly forwards the argument of the individual fish as an individual sentient body. Animal rights positions tend to be the most uncompromising against recreational fishing. However, as I hope I have made clear at this point, this position is trapped in an anthropomorphic logic that equates all life as equivalent to human life, including the phenomenological responses of each form of life to surrounding conditions and experiences. Ideological positions drive this belief rather than concrete evidence. Animal rights positions, that is, stem from gospel rather than data. However, as responsible anglers, we should not dismiss these positions out of hand but should instead consider how such philosophies might inform our catch-and-release practices to better reduce the infliction of pain.

The critical position frequently taken up against recreational angling contends that the very idea of *recreation* suggests that anglers are *playing* with sentient beings or with potential food resources for no reason other than pleasure, leaving in question the ethical value of recreational angling and the intrinsic value of play. As Arlinghaus and Schwab explain, "the objection is quite common in informal context, and the critique claims that what is wrong is that recreational fishing is carried out for fun, largely irrespective of what a fish might feel."[17] For clarification, we should contrast the idea of fishing "for fun" with fishing for subsistence, which is construed as an ethically "legitimate" reason for causing fish pain and harvesting those fish. That legitimacy relies on an unwavering belief in a foundational truth that the use-value of fish as food is both given and constant—and, as I will explain later, that providing for an anthropocentric protein economy will always be assumed not just as ethical but as naturally necessary. Even when framed in an animal welfare context, harvest for food preexists as an acceptable use for fish. Popular opinion tends to understand fishing for essential-need harvest as permissible, whereas fishing for recreation is questionable from oversimplified ethical positions. In fact, philosophically, fishing for food—that is, identifying the recreational dynamic as a relationship between an agent and an object—naturalizes the act of the agent taking the object for the intent of consumption as acceptable within a natural order of things. This long-circulated narrative that defines such an act as an inherent part of the human condition, or an inherent part of the "natural" relationship between human and nonhuman, preordains justification for the harvest mentality as ethically superior to catch and release, despite the repositioning of conservation and sustainability as ethically favorable over any form of killing. As Balaz might explain it, if you are going to kill it, you've got to eat it, because that is the only ethical response to the act of the harvest. Thus, we can say that part of the ethical question surrounding recreational angling is a question of motive and of how such motives have been portrayed historically as natural or unnatural, as interaction between natural entities or as technological conquest. Arbitrary slaughter, as I described on the Skyway Pier and other fishing piers, ought to be rendered unethical no matter the justification. Unfortunately, though, it is deemed culturally acceptable among many anglers because of an underthought, selfish use-value declaration: they are of no use to *me*.

More concretely, though, we should understand these ethical dilemmas as playing out most fundamentally at the level of the individual, at the level of the angler's intentions and actions and how he or she interacts with the marine environment and its inhabitants, philosophically and actively. Yet the multidimensional characteristics of what might constitute "recreational" fishing for any individual angler are likely to vary from those of another; thus, the very notion of what constitutes "recreation" is accessible only at the level of the individual. For example, as Arlinghaus and Schwab put it, if an angler catches and kills a fish, and the kill is "eaten, and thus the intention of the angler was to engage in a natural act of predation concomitant with a range of positive emotions, there seems to be no reasonable ethical objection."[18] On the other hand, some—more often than not, nonanglers—identify catch-and-release fishing as disrespectful to the fish because of the potential infliction of pain on the fish. Death for consumption trumps pain and survival in an unwritten hierarchy of what is right. That is, killing and eating is understood as more ethical than catch and release in this context. Keep in mind, too, that bound up in this distinction is a subset of intent: that of aggression. Anglers tend not to enter into recreational harvest with an aggressive intent to cause harm (with the possible exceptions of the teenage clan I described earlier and others lacking empathy). This distinction is important because if aggression were a motivational facet of angling harvest, then the charge of causing pain would need to be read as intent to cause pain, a much more serious charge. However, if we are to attribute the recreational angling ethic not to an outside observer who can foist an ethic or a critique upon the angler but to the angler, then more crucial is the intent of the angler rather than the external reading of the situation by one not directly participating in the activity. That is, the ethic should be driven by the participant's intent and action, as the ethic of most catch-and-release anglers is. To put it another way, ultimately, the individual angler's ethic should be determined not by the perceptions of nonparticipants but by the shared values of the participant community.

There is, though, at least one philosophical impediment to

my assertion of ethical authority, which cannot be accounted for with any degree of simplicity. These are complicated and complex issues that deserve at least our acknowledgment if not our consideration and respect. First, the new anglers' ethic must acknowledge that fish are a part of the commons and that those who do not participate in recreational angling or commercial harvest have some degree of claim over them as resources of the commons. As I noted earlier, any fish I catch is not my fish. It is, in fact, as much the nonangler's fish as it is the angler's. To be clear, though, that does not authorize the nonparticipant to determine the harvest ethic of the partici-pant. It only grants the nonparticipant the right to become a participant. Likewise, we must also consider whether the non-participant has the ethical right to assign a proxy participant or whether others can claim a role as proxy participant without nonparticipant ratification, as is the case with commercial and industrial harvest, which I take up in the next chapter.

How, then, do we negotiate the complexities of anglers' ethics in the twenty-first century? First, such ethics must originate with anglers and with a deliberate willingness to talk about ethical approaches to harvest and catch and release. Such dialogues will be uncomfortable—likely hostile, if we are willing to admit it—but necessary. If we are not transparent in our desire for and con-struction of anglers' ethics, we risk having policies imposed on our angling rights in the guise of ethics. Only by establishing flexible and condition-malleable ethical guidelines informed by scientific data will we ever be able to maintain a culture of ethical angling that will be able to work to sustain the world's ocean while promoting the cultural value of recreational salt-water angling.

Second, all discussion of anglers' ethics—no matter how criticized—must account for individual angling actions within the scope of global ocean conservation. That is, we must acknowledge that any harvest we engage has a larger, global effect. This may be a difficult concept to come to terms with at the individual level, since it is likely that any individual angler's harvest is insignificant in relation to global fisheries or ocean conservation. However, given the vast number of recreational anglers now fishing in the United States, we cannot dismiss the individual angler's impact, because the impact of that individual angler is compiled with that of every other individual angler. The individual angler must be accounted for as a member of a collective. Thus, the individual angler's minimal harvest is compounded across the recreational angler population, having a noticeable effect on not just fishery populations but on every aspect of human engagement with the ocean. We may fish alone, even seek the solitude of the angler's devotion, but our actions aggregate.

Ultimately (and perhaps realistically), it is unlikely that anglers will devise and reach an ethical consensus. The diverse backgrounds of anglers, angling approaches, geo-cultural practices, and commonplace beliefs render consensus implausible, not to mention the complex range of local and global contexts that may influence local practices and individual ethics. Despite such intricacies, we find ourselves at a historic moment when it becomes prudent to hold in reserve our traditions of rugged, frontier-forged individualism in favor of a confederation of like-minded anglers willing to advocate and hold true a recreational anglers' ethic that accounts for the needs of recreational anglers and the ocean, both present and yet to come. In addition, let me emphasize that to do so requires advocacy. To take up the possibility of such ethical approaches, though, the recreational angler, individually and collectively, must also account for the ideologies, philosophies, legislations, and cultures that conflict with the recreational anglers' ethic and that compete for claim over the use-value of the ocean and its inhabitants, and advocate on behalf of the recreational anglers' ethic.

Such a position may appear lofty, but anglers' ethical adjustments can begin—should begin, in fact—more simply and locally with the individual angler. For example, to better understand my own fishing and to better situate my fishing in larger ethical frameworks, I have begun to conduct a simple angler's self-assessment from time to time. Curiosity drives this assessment or audit, but its results often contribute to my overall thinking about how I fish and why I fish. I do not make this assessment every time I fish, but the more often I do, the more ingrained the thinking becomes in my fishing.

Here is how it works: Let us imagine that I have just spent the day fishing, and it has been a fairly slow, unimpressive day. During the day, I managed to catch one ladyfish, which I kept for cut bait; three undersized speckled trout, which I released; and one slot-sized redfish, which I kept for table fare. With this in mind, I begin to ask questions in two categories: biological ecology and material ecology.

BIOLOGICAL ECOLOGY

1. How many animals did I kill today?
Knowingly, I killed the ladyfish and the redfish. I assume that I did not kill the trout that I released because I released them with appropriate precautions, and research shows that trout have a low mortality rate after release; they are resilient. I also killed about thirty shrimp, which I used as bait during the day. I did not accidentally catch any other fish that might be regarded as trash fish or bycatch.

2. Where did each of the animals I killed fit in the local ecology?

By killing the ladyfish, a common prey fish for many species, I removed that protein and energy from the ecology, denying it to another predator. Likewise, the ladyfish is also a predator that eats smaller fish and crustaceans. By removing it, I decreased the amount of protein and energy removed from the system by the ladyfish. By removing the redfish, I removed a significant portion of protein from the oceanic ecology and redistributed it into my energy system by eating it. (I may even note that by throwing scraps such as skin and bones into a trash bag rather than back into the ocean, I removed that protein from where it might otherwise have remained part of the energy cycle.)

MATERIAL ECOLOGY

1. How much fuel did I burn today?

Between trailering the boat and running the boat, how many gallons of fuel did I burn? What is the cost of that fuel? What footprint does burning that much fuel leave?

2. How many plastic lures did I lose to short bites and break offs?

3. How much line did I dispose of or lose?

4. How much plastic did I depend on today for sandwich wrappers, drink bottles, lures, and other components of my tackle?

Our new anglers' ethic will depend on our willingness to repeatedly ask these kinds of self-reflexive questions, to continually ask about and confirm our understanding of our individual ethical positions.

5

Sea Lords, or The Rhetoric of Sustainable Seafood

Americans don't own the fish in their oceans anymore, not really.

Lee van der Voo, *The Fish Market*

Ben Raines made the term "Sea Lords" popular. As he has explained, he adapted it from federal fisheries managers who use it to delineate a group of people who make their living by leasing commercial harvest rights to other fishermen much as a landlord might rent housing space. The term, of course, is appropriated from the Royal Navy, where the designation refers to the admiralty—quite literally, the Lords of the Sea, referring to those historically entitled by lordships to govern the sea—the First Sea Lord now being the Chief of Naval Staff. However, for Raines, the term reflects an even more tyrannical aspect of lordship, with titles bequeathed by the government to a few not to govern the sea—a concept that historically carries the weight of stewardship—but to harvest its resources for sale and profit. Moreover, as Raines points out, where there are lords there are serfs, whose labor profits the lords. For Raines, the title "Sea Lords" is emblematic of an unjust system of fisheries management that, though often touted as environmentally sound, prevents access to fisheries resources to all but the Sea Lords through a system commonly known as "catch shares," arguably the most problematic approach to fisheries management in human history. To be reductive about it, though entirely accurate, catch-share policies seek to privatize the world's ocean and fisheries.

Raines is a reporter for AL.com, an online media amalgam of the *Birmingham News,* the *Huntsville Times,* Mobile's *Press-Register,* and the *Mississippi Press.* Over a four-month investigation, Raines found that more than $60 million had been earned by a handful of catch-share owners who employed fishermen—laborers who are required by law to turn over more than half of the profits they earn from fishing to the catch-share owners. Officially known as the Individual Fishing Quota (IFQ) system, catch shares allocate ownership of harvestable fisheries to a few owners. Caught in the wake of catch-share policies are myriad anglers, businesses, and communities, all paying for access to or denied rights to fish. Wrapped in greenwashing rhetoric, catch-share advocates defend the policies as sustainable approaches to fisheries management, and in some ways, they may very well be. However, the catch-share model embraces a particular kind of sustainability at the expense of democratic access—including access by recreational anglers.

Catch-share[1] policies work this way: the federal government—in the form of NOAA and the eight regional fishery management councils—assigns to those who qualify (these can be individuals, corporations, or even boats) the rights to harvest fish in a specific region. The same governing boards determine how many fish those given the rights might harvest. Those rights are initially dispensed based on an entity's previous record in harvesting fish when the catch shares are put in place. For example, if a catch-share policy were initiated this year in a specific region, those shares would be distributed based on the amount of fish an entity harvested the year before the policy was established. Therefore, if you harvested a significantly larger number of fish than did your competition prior to the catch-share policy implementation, you would be granted a significantly larger portion of shares, while your competitor would receive fewer shares or none at all. Neither the conditions and context of your previous harvest nor those of your competitors would have any bearing on the allocation. Therefore, if your competitor missed a substantial part of the harvest season because of equipment failure or health issues, or if he or she simply had a bad harvest compared with other years, those details would not be accounted for in the share distribution. Once the governing agency distributes the shares, the recipient of those shares owns them. Owns. Redistribution is not revisited each year, or even later. The shares, and thereby the exclusive right to harvest the share of fish, remain the confirmed property of the owner. Shares may then be bought or sold, as might any other commodity. Of course, as with any valuable commodity, there are those determined to own as many shares as possible in order to control the industry.

Share owners are not required to harvest the fish to which they have rights; they may hire others to harvest in their behalf, and most do. Thus, many of the hardworking American anglers who fish commercially do so to serve someone else's profit. Likewise, because shares may be bought or sold, a few share-holders have been buying up rights to as many shares as they can, gaining control over entire fisheries. Because those without shares are not permitted to harvest, small businesses, individual anglers, and family businesses find themselves strangled out by multishare owners and large corporations. Technically, the federal government could take back the already-distributed catch shares and reapportion them, but in the last quarter century since the implementation of catch shares, it has not done so, allowing fewer and fewer organizations and individuals to own more monopolizing shares of fish populations. These policies, then, not only bar access to fish but unequivocally deny people the right to fish. In doing so, catch shares not only privatize America's fish stocks, remanding them to catch-share owners, but also eschew citizen rights.

Catch-share policies are one of the most critical aspects of how we think about and act regarding the future of the world's ocean. Catch shares affect not only commercial harvest approaches but recreational harvesting, as well. Because federal agencies create all fishing policy—recreational, commercial, and industrial—all federal policy has entwined all types of harvest into one chaotic, writhing—and drastically misman-aged—Gordian policy knot. In 2016, for example, the Gulf of Mexico Fishery Management Council, in determining catch-share allocations, allotted 51 percent of the total harvestable quota of American red snapper to commercial catch-share own-ers, leaving 49 percent of the quota to recreational anglers, and subsequently dividing that harvest between licensed for-hire anglers (guides) and private (recreational) anglers. Likewise, the council dictated that the recreational harvest season in federal waters would last a total of nine days out of the year, restricting when recreational anglers might be able to harvest. A report by the Southeast Regional Office Limited Access Privilege Program (SERO-LAPP) titled *2016 Gulf of Mexico Red Snapper Recreational Season Length Estimates* summarizes and explains (or obfuscates, depending on how you read it):

> Median projected federal season lengths from the ten pro-jection scenarios for 2016 were 48 days for for-hire (range: 38–56 days) and 8 days for private angler mode (range: 6–9 days). Based on the most recent two years, the mean season length would be 46 days for federal for-hire mode

and 9 days for private angler mode. Across the ten projection scenarios, a median of 21 percent of bootstrapped private angler model runs and a median of 17 percent of bootstrapped federal for-hire model runs exceeded the ACL under these season lengths. The May 2, 2016 SERO-LAPP-2016–04 state water seasons established by each state in 2016 are projected to land approximately half of the private angler component ACT and reduce the median private angler federal season length from 15 days to 8 days. Assuming inconsistent state seasons, the implementation of RF-40 reduced the private angler mode federal season length by 8 days (a 50 percent reduction) but increased the federal for-hire season by 32 days (a threefold increase). Bootstrapping was used to capture model uncertainty; assuming sector separation and inconsistent state seasons in 2016, the mean risk of exceeding the ACL at a 20 percent ACT buffer was estimated at 15 percent for federal for-hire and 17 percent for private angler mode.[2]

Bound up in a sixty-eight-page report justifying recreational harvest limits, this brief though turgid paragraph reveals an implicit mind-set that considers recreational fishing management an afterthought to establishing commercial harvest priorities. Thus, though neither "catch shares" nor "IFQ" appears anywhere in the report, the marrow of their significance is felt as a prevailing influence. The mismanagement of red snapper fisheries deserves substantially more attention than I grant here, but I bring it up simply to identify the ways in which catch shares directly affect recreational fishing policies and the inseparability of conversations about commercial harvest and recreational harvest policies.

It is not merely fisheries management that is encumbered
in catch-share policies; there are significant philosophical and
ethical issues at play here that require the attention—and
advocacy—of the recreational angling community and that
have direct impact on the future of the world's ocean. Most
obvious, of course, is the very idea of the privatization of the
world's ocean. Let me be clear here, too—many catch-share
supporters now seek to bring the idea of catch shares to other
countries, hoping to obtain exclusive harvest rights in foreign
waters as well as domestic. Chief among these is the Walton
Family Foundation, heir to Sam Walton's Walmart empire.
Veiled behind greenwashed environmental rhetoric professed
by the Environmental Defense Fund (EDF), Clan Walton and
other catch-share advocates have begun pressing the European
Union to impose catch-share programs, as well. I will come
back to this in a bit, as it, like the overall catch-share issue, is a
confounding and complex situation.

Oregon-based journalist Lee van der Voo's *The Fish Market: Inside the Big Money Battle for the Ocean and Your Dinner
Plate* is the most comprehensive chronicle of the evolution of
catch-share programs. Early in the book, she clarifies three key
aspects about catch shares:

> First, the private property rights attending catch shares
> had locked many fishermen and even whole communi-
> ties out of the oceans. Second, catch shares had created
> powerful landlords on water. And third, as those landlords
> grew more powerful, catch shares were converting fisher-
> men from proud family-business sorts into sharecroppers
> who were leasing their access to the sea from wealthy and
> increasingly corporate power brokers.[3]

Ultimately, catch-share policies are founded in economic philosophies, not ecological, not conservationist, not even environmentalist philosophies, as proponents often claim. They are about making money. Granted, as sound capitalists, we might say that making money is a worthy endeavor; however, catch shares go a step beyond and regulate who can legally make that money—and thereby who cannot, thus eliminating any sense of capitalist competition. There are many residual effects of this kind of policy making; most evident are the effects on the economy and on the independent commercial fisherman and businessman. More complicated are the effects of catch shares on recreational anglers and on ocean conservation. In order to unpack this relationship, recreational anglers need a better understanding of the relationships at play in catch-share policies. In fact, as noted in the previous chapter, a comprehensive recreational anglers' ethic requires advocacy, which demands critical understanding of the policies that affect our rights and privileges as anglers. Our "rights" must embody a culture of conservation and sustainability.

To simplify, catch-share programs work this way: each of NOAA Fisheries' eight fishery management councils determines the harvest allocations for fisheries in its region. These allocations include the length of the harvest season, how many fish can be harvested in a given season or year, and how many of those fish can be harvested by commercial anglers and how many by recreational anglers. For example, the Gulf of Mexico Fishery Management Council determines how many red snapper should be harvested in a given season. This council then divides the allocation between commercial anglers and recreational anglers; traditionally, this division has favored commercial harvesters, even if minimally, with splits such as 48 percent to recreational anglers and 52 percent to commercial harvesters.

In regions where catch-share programs have been established, the regional council also determines which of the commercial harvesters are permitted to harvest fish and what percentage of the overall take each approved harvester is allotted. That is, the council subdivides the commercial shares among a select number of commercial fishermen, limiting how many commercial fishermen can participate in the harvest and regulating what percentage of the catch each can take. That is, catch shares eliminate competition and assign legal ownership of fishery shares to a narrow field of harvesters.

The allocations of commercial shares are determined, divided, and distributed by the council based on a tally of fish caught in the year prior to the assignment of shares. That is, when a council determines which commercial entities should receive what percentage of the shares, it looks at how many fish a commercial harvester took the year before, rewarding those with higher catch numbers with greater shares. Those with low catch numbers are frequently denied shares altogether, thus legally barring them from continuing to work in the commercial fishing industry. In regions where catch-share allocations have been implemented, such as the Gulf region, these allocations were made once and have been kept in place ever since. Shares are not determined annually, as one might expect. There is never a review or a reallocation. In essence, those who have been granted shares by the councils have been handed perpetual legal rights to ownership of a fishery without having to compete for that fishery. That is, the government has bestowed exclusive ownership of a public resource on those harvesters.

Think about it this way: imagine a birthday party when it is time to cut the cake. We anticipate that the cake will be cut into enough pieces for everyone to have a piece. There may even be a few leftover pieces. However, the host of the party steps forward and announces that first she is going to cut the cake into two pieces, one slightly bigger than the other. With the larger piece, she asks how many people at the party attended the same party last year. Ten people raise their hands. The host asks how many pieces of cake each ate last year. Two say they ate three pieces each; three say they ate two pieces each; and five say they each ate one. With this information, the host then cuts the larger portion of the cake into twelve pieces, giving three pieces each to the two who ate three pieces last year and two pieces each to the three who ate two pieces. Those who ate only one piece the previous year are told that they will not be getting cake. The hostess then informs the guests that these same people will receive the same share of cake next year and in all subsequent years, and that those who did not receive any cake will not receive any in subsequent years, either.

To complicate matters, catch-share owners can buy and sell their shares as they see fit. Thus, many catch-share owners with large shares actively seek to purchase shares from others holding smaller shares, increasing their own holdings. Some shareholders work to acquire as many shares as possible within a given fishery, in essence monopolizing the fishery. Chief supporters of such practices are members of the Walton family and the Walton Foundation, heirs to the Sam Walton Walmart fortune, the largest corporation in the world led by one of the wealthiest families in the world. Because catch-share owners are not required to harvest their shares themselves, they can lease their shares to others. Thus, like feudal lords, these Sea Lords lease their shares to those eager—if not desperate—to work the sea, in essence turning those who harvest the fish into sharecroppers. And of course, share owners lease harvest rights at rates designed for their own profit, leaving little for the laborers. Van der Voo, Raines, and others report numerous cases of fishermen who lease harvest rights struggling financially while share owners profit. High lease rates and low profits for fishermen not only make it difficult for the leaser to make a living but also make it impossible for the leaser to ever be able to purchase shares of his or her own. I should note, too, that there is substantial anger in the catch-share debates regarding a few people who own substantial shares of US fisheries who have never even set foot on a fishing boat and have never participated in the act of catching those fish. There is a cultural understanding among commercial harvesters that those who profit from the harvest ought to be the ones who actually do the work of the harvest. Owners who don't fish and who subsequently rob fishermen of the rights to fish are not viewed favorably by those who have traditionally participated in commercial harvest.

Perhaps more problematic, though, is that the fisheries that NOAA councils allocate to share owners are part of the US commons, communal property of the citizens of the United States. However, the catch-share programs assign private ownership of those elements of the commons to a select few. Unlike with lease programs, such as those used when public forests are leased to lumber companies or mineral rights on public lands are leased to oil companies, catch-share owners pay no taxes, pay no leases, pay no fee back to the people of the United States. They contribute nothing to the restoration of those resources harvested, as leasers do in the oil and timber industries. The US government freely gives these fisheries to a small number of private entities and declares them legal owners. To this day, there has been no review of allocations in any catch-share program, no attempt to reallocate shares, no attempt to spread the wealth among others, no attempt to evaluate the success or failure of the catch-share program in terms of either financial or conservation success.

Frighteningly, most Americans have no idea that the fish
they eat, which are harvested from the waters they collectively
own, have been deemed the private property of a few Sea Lords
who sell those citizens' own property back to them in frozen
filets. The recreational angler might be tempted to dismiss
the catch-share fight by believing that it is a problem for the
commercial part of the allocation; we still have our recreational
allocation. However, as catch-share owners gain stronger
foothold with the NOAA councils, we see allocations leaning
more heavily toward the commercial side of things, while recreational allocations risk being reduced. As van der Voo explains
regarding both commercial and recreational angling on the US
West Coast, "90 percent of fish that can be caught on the West
Coast belong to trawl boats."[4] Likewise, fisheries management
conversations have also begun considering the possibility of
imposing catch-share programs on charter captains and guides,
limiting the guides who might be able to work in a given
region.

What may be most frightening—though rhetorically
effective—is the alliance catch-share advocates have formed
with environmentalists, or at least with the appearance of
environmentalists, and the use of environmentalist rhetoric
to defend the catch-share program. One of the largest
supporters of the federally mandated catch-share program is
the Environmental Defense Fund (EDF), a nonprofit environmental advocacy group. The EDF, which began advocacy
work in the late 1960s by using Rachel Carson's research in
Silent Spring to successfully initiate a ban on DDT in Suffolk
County on Long Island, New York, to protect resident osprey,
has become one of the most powerful environmentalist lobby
groups in the country. It weds science to economics to forward its agenda and employs this strategy as the most vocal

supporter of the catch-share program. However, despite efforts to appear bipartisan and untethered to corporate influence, the EDF receives substantial funding from corporate groups such as Walmart and McDonald's, both of which have substantial investment in the catch-share debate.

The Walton Foundation, in fact, donates tens of millions of dollars to the EDF for its work in ocean and fisheries lobbying. The EDF in turn has paid substantial money to initiate the Gulf of Mexico Reef Fish Shareholders' Alliance, a nonprofit organization that represents the interests of commercial fishermen. Through the Shareholders' Alliance and other lobbyist groups, the EDF has been able to push a catch-share agenda, working to commodify the nation's fisheries. Part of the motivation for the Walton Foundation in funding such efforts can readily be seen in the relationship between the Waltons' support of catch-share programs and the amount of fish products sold in their stores. As van der Voo puts it, the Walton Foundation believes that "nature is forever at risk without financial rewards linked to preserving it."[5] Again, we see the catch-share program developing from economic rather than conservationist philosophies. No matter the motivation, though, the Walton Foundation is one of the two largest contributors to developing and maintaining catch-share programs. Along with the Gordon and Betty Moore Foundation, the Walton Foundation has donated about $10 million in support of catch-share programs. As an example of the intimacy between the Waltons and the EDF, the EDF maintains one of its eight offices in Bentonville, Arkansas, adjacent to the Walmart home offices.

The EDF supported many environmentalist agendas prior to its relationship with the Waltons, and certainly before catch shares emerged. The EDF is quick to note that it began its work in catch-share programs in response to extensive overharvesting in US waters. It hoped to galvanize many of the disparate entities affected by overharvesting. In 2000, the EDF established a policy mandating that its work in ocean protection be linked fundamentally to economic strategies, confirming an institutional understanding of ocean as resource. This policy began to affiliate other groups with the EDF to address overfishing. For example, the EDF is a member of the Marine Fish Conservation Network (MFCN), a group formed in the 1990s with a mission to end overfishing, which claims allegiance between commercial, recreational, scientific, and citizen communities. The MFCN works to ensure that the Magnuson-Stevens Act (MSA) is maintained in its present form. However, the MSA—which is written from the perspective of commercial harvesting and applies commercial standards to all harvesting, including recreational angling—has now been challenged by recreational anglers. That is, both the EDF and MFCN work to maintain status quo policies established by the MSA in order to maintain catch-share rights of harvest exclusivity, and in turn the exclusion of all other anglers—commercial, industrial, and recreational.

Fortunately, even under a banner decrying overfishing, the EDF found that the commercial interest as a whole did not support catch shares. While the EDF argued that fewer boats fishing would abate overfishing pressures, many commercial fishermen saw the strategy as a way to exclude them from harvest rights. Even those commercial interests that saw value in the conservation logic recognized the impact of the exclusivity and called for catch shares to rotate and renew. The EDF adamantly supported private property rights as the only avenue for ocean conservation, pitching catch shares as a panacea for overharvest.

However, what the EDF did not address was the economic toll its catholicon had on those not granted shares. Van der Voo points to catch-share policies implemented in Alaska that eliminated jobs and "instituted a renting class."[6] She also compares the situation to the socioeconomic pitfalls of North Carolina's tobacco quotas, which established an impoverished sharecropper population—a problem, van der Voo notes, that took more than a decade to remedy through farm bill legislation.[7]

In response to dissent from within the MFCN, the EDF withdrew from the MFCN and increased its pressure on federal policy makers, virtually demanding that private property rights be imposed on fisheries. Without the consensus support it had hoped to gain from the MFCN, the EDF turned to the Charles Koch Foundation, the Reason Foundation (a libertarian-leaning policy advocacy group), and the Property and Environment Research Center (which is dedicated to "improving environmental quality through property rights and markets").[8] Along with the Walton Foundation and the Gordon and Betty Moore Foundation, the EDF has promoted its "environmentalist" lobbies through some of the most conservative supporters around. Van der Voo notes, too, that prior to 2003 the EDF budgeted $2.5 million for its ocean programs, but with support from these organizations, these programs now top $24 million.[9] This funding supports scientific research that defends catch-share programs, validating the success of those already granted catch-share rights. However, as van der Voo so rightly puts it, "the benefits of preassigning the catch, and limiting how many fish could be caught based on science, were being conflated with the notion of privatizing the sea." Grounding its rhetoric in science, though, gives the EDF the appearance of adhering to a strictly scientific, unbiased approach to fisheries management. Yet, as van der Voo explains, Timothy Essington, associate director of the School of Aquatic and Fishery Sciences at the University of Washington, reviewed the data attached to catch-share implementation and found that catch shares resulted in "no significant ecological benefit." According to van der Voo, the only benefit Essington identified as the result of catch-share implementation was the "ability to make fisheries more predictable and fish more valuable."[10]

Again, the issue of value is remanded to economic value, not ecological or cultural value. Economic value is powerful in policy determination. In places where catch shares had been implemented, such as Alaska, catch-share owners were certainly profiting (van der Voo notes that in Alaska between 1992 and 2009, commercial interests earned $234 million),[11] but the actual harvest rights had been consolidated to fewer and fewer owners, resulting in a growing disadvantaged population dependent on fishing for their living.

As I have noted, the Magnuson-Stevens Act has been amended a number of times since its inception in 1976. In 2006, one such revision enforced hard harvest caps on the numbers of fish that fishermen could take from US waters. These caps, the amendment indicated, were to be based on specific scientific data. The revision, conservationists assumed, would restrict overall harvesting; however, what the language did was support the EDF's rhetoric of science-based control of who could harvest, essentially validating catch shares as an inherent part of US fisheries management. "It was a move," van der Voo writes, "that underscores how 'overfishing' had really been 'under-regulating.'"[12] The EDF's rhetoric was naturalized into the conversation, giving it significant power in the overall debates. Likewise, given the funding the EDF received to promote catch shares, its voice began to silence any opposition.

In fact, in March 2017, the South Atlantic Fishery Management Council held a public meeting on Jekyll Island, Georgia, to address a proposed catch-share program. In collaboration, four commercial fishing interests, operating under the name South Atlantic Commercial Fishing Collaborative, filed a request for an Exempted Fishing Permit (EFP—the official name for catch shares). The permit would grant rights to twenty-five commercial boats to harvest gag grouper, vermilion snapper, blueline tilefish, greater amberjack, gray triggerfish, and a variety of jacks for a two-year trial period while being exempt from standard fishing regulations. The pilot program would essentially provide ownership of these fish species to the twenty-five boats with the opportunity to turn the pilot program into a permanent permit. The intent of the petition, the collaborative claims, is to move its snapper-grouper fishing away from competitive derby-style fishing in which all boats compete for a limited number of fish, a method it claims creates quota averages, numerous fish discards, and marketplace instability that manifests in fluctuating prices.

Many of us saw the proposal as a brazen attempt to take control of South Atlantic fisheries using unethical approaches toward an unethical end. Key among the problems was that the collaborative included two current, sitting South Atlantic Fishery Management Council members and one former member, the very people who would vote on whether to approve the petition. That is to say, at least two of the council members stood to profit from the approval of the permit. Prior to the meeting, the council did not address any conflict of interest or breach of ethics in allowing the proposal to be heard.

Likewise, the collaborative very carefully constructed the proposal to avoid any language that might link the request to the growing resistance to catch shares, avoiding phrases such as "catch shares" or "Individual Fishing Quota." Instead, the proposal referred to "output-based systems" and "allocation-based systems," ambiguous terms used to obfuscate the proposal's ultimate objectives. Nor did the proposal identify how the shares would be allocated or traded.

Writing in *Sport Fishing*, one of the most widely read recreational fishing magazines in the country, Jeff Angers, president of the Center for Sportfishing Policy, decried the proposal: "This EFP is like others that have been rubber-stamped through the federal system—it ignores all the thorny issues and presents a bully pulpit for these lucky 25 commercial vessels to tout how wonderful catch shares work for them so that it will eventually be accepted as a fait accompli. If it is anything like previous EFPs, we will eventually find out that those lucky 25 vessels all happen to be owned by just a handful of people."[13]

Although organizations such as Saving Seafood, a nonprofit corporation funded by the commercial fishing industry that serves as a public outreach arm for that industry, encouraged supporters to attend the Jekyll Island meeting to voice support for the EFP proposal, the collaborative retracted its proposal after the council received significant public outcry against the permit. Leading the charge against the proposal, North Carolina representative Walter Jones (NC-3) wrote to the council citing widespread opposition to the proposal, noting in particular the numbers of oppositional comments posted to the council's online comment page when it requested comments in response to the draft of the proposal. Jones also noted that the council had rejected a similar proposal in 2013. Jones wrote: "Clearly, the overwhelming sentiment of permit holders, and the precedent of prior council actions, argues for opposition to this application. The snapper-grouper permit holders of the South Atlantic have not only not given their consent to this proposal, they appear to be nearly unanimous in their opposition."[14]

Paramount, too, in the resistance to the proposal was the

voice of the North Carolina Fisheries Association, a group dedicated to protecting the traditions of commercial harvest in North Carolina. This group has long opposed catch shares, seeing them as putting small commercial fishermen at risk of losing their businesses.

The retraction of the South Atlantic Commercial Fishing Collaborative proposal was not touted as a victory against catch shares. Instead, Angers and others saw the move as merely a delay before the collaborative makes a stronger push for the permit. That push, we all assume, will include increased pressure from the EDF. Angers, and lobbyists from the American Sportfishing Association, Center for Sportfishing Policy, Coastal Conservation Association, Congressional Sportsmen's Foundation, National Marine Manufacturers Association, and Theodore Roosevelt Conservation Partnership, agree that the problem is that at other council meetings across the country, the EDF has been able to stack the deck in its favor. The EDF, given its massive funding support from the Walton Foundation and the Gordon and Betty Moore Foundation, regularly pays people to attend council meetings when public comments are called for. Those opposed to catch-share programs—often underpaid commercial fishermen or everyday recreational anglers—are unable to travel to attend the meetings. Travel costs and time away from work impede their ability to participate in the process. Yet the EDF can fund supporters to attend. Several anglers have reported to me that when they attend such meetings, they hear EDF-paid supporters joking about how they receive a stipend as well as compensation for travel, lodging, and food for attending. It is tough, they tell me, to argue against a proposal with only a handful of opposition voices in attendance when the conversation is dominated by paid influencers in support of such proposals.

In the months that followed the retraction of the EFP proposal, I turned my attention back to the water, my mind focused on the titillation of late spring redfish and snook and the anticipation of warming waters offshore. My tackle bag was full of new lures I planned to review for InventiveFishing.com, and such work had me as attuned to weather and water conditions as to keeping up with federal fisheries policies. On the water, I drifted between thoughts of migrating fish and the human arguments about who could lay claim to those fish. The fish, of course, were unaware that their lives were being debated as commodity and parceled out in shares.

On boat ramps, in tackle shops, and in other places where recreational anglers tend to gather, I joined friendly conversations about fishing. Such impromptu discussions often turned to complaints about regulations: unjustified short seasons for red snapper, shifting regulations for redfish, undermonitored catches of speckled trout, and so on. Yet in none of these friendly exchanges did I hear mention of the role of catch shares or federal policies that would ultimately affect the very issues anglers around the country were facing on all coasts. I became more and more aware that those dedicated to the saltwater fishing life were often uninformed about the political forces that affected their own fishing lives. I began to ask in such conversations whether others knew about catch-share programs or about revisions to the Magnuson-Stevens Act, or why the very regulations they grumbled about had been implemented. Few did.

Standing with the world's ocean before us, we find ourselves in a precarious moment. Our ocean faces desecration from many provenances. Our awareness and involvement in the nation's ocean management requires that recreational anglers be alert to the decisions legislators make affecting all fishing, not just fishing of local concern. Eleven million saltwater anglers have the potential to be a significant voice in how such policies are addressed.

The problem is not exclusively a US issue, as catch-share logic has begun to take hold globally. In New Zealand, for example, 86 percent of harvest rights are owned by twelve fishing companies. In Iceland, over a four-year period, twenty companies were able to take control of 66 percent of all fishing rights.[15] Groups such as the EDF, and the Walton Foundation specifically, have also begun to introduce catch-share thinking to other countries and foreign governing bodies such as the European Union, hoping to expand their hold on the world's fisheries.

If recreational anglers—both saltwater and freshwater—are to understand their role in ocean conservation, then we must acknowledge conflict with oppositional views about resource management. At the heart of such conflict lies the ideological position of identifying fish and ocean as simply a resource that can be portioned out to private owners. To anoint fish as a resource, as food or oil to be owned, harvested, and sold, devoid of any ecological understanding of the relationship between human, fish, and ocean, stands as detrimental to the ocean's future.

6

Fish Sandwich

This is not the first time humanity has glanced across
the disorderly range of untamed nature and selected
a handful of species to exploit and propagate. Out
of all of the many mammals that roamed the earth
before the last ice age, our forebears selected four—
cows, pigs, sheep, and goats—to be their principal meats. Out of all the many birds that darkened
the primeval skies, humans chose four—chickens,
turkeys, ducks, and geese—to be their poultry. But
today, as we evaluate and parse fish in the next
great selection and try to figure out which ones will
be our principals, we find ourselves with a more
complex set of decisions before us.

Paul Greenberg, *Four Fish: The Future of the Last Wild Food*

A substantial part of the problem with the catch-share argument is the issue of access and ownership. Catch-share proponents argue that controlled commercial harvest provides access
to seafood to American citizens who would otherwise not have
avenue to what rightfully belongs to them as much as it does
to those who live near the ocean. Catch shares, they contend,
ensure a consistent supply of food fish to noncoastal regions.
Commercial harvesters, the argument goes, serve as agents
for American citizens, providing a public service (for a price)
in delivering seafood to noncoastal residents and businesses.
Catch shares, then, garner support from restaurateurs, seafood

retailers, and seafood consumers nationally who do not or cannot catch their own fish. In the United States, seafood from around the world is as commonplace as beef, chicken, and pork, and US consumers expect it to be delivered on par with farm-raised, privately owned livestock—suggesting, too, a mind-set that understands "seafood" to equate with farmed meat.

To meet such demands, not just in the United States but globally, harvest numbers continue to increase. As a result, overfishing has become one of the most pressing issues not just for sustainably conserving the ocean but also for addressing poverty and nourishment in underdeveloped nations. The United Nations identifies overfishing and global poverty as inextricably linked. According to the UN's Food and Agriculture Organization (FAO), 5 percent of the world's gross domestic product depends on marine resources, predominantly fish. Likewise, the UN shows that 10 percent of the global population depends on the fishery industry either directly or indirectly for its livelihood. However, these numbers hold true only in official measurements. Illegal, unreported, and unregulated fishing add to the detrimental costs of overharvest. According to the FAO, the United States alone loses an estimated $23 billion per year from illegally harvested fish. Of course, that accounts for solely an economic loss, not an ecological loss, as economics is the prevailing measure by which we assign ultimate value. The ecological impact of overharvest and illegal fishing is much more significant than just the lost dollar value. In May 2017, the Indonesian news agency Antara identified that more than 30 percent of the global loss in fisheries harvest revenue can be traced to illegal fishing originating in Indonesia. In 2014, Indonesia appointed Susi Pudjiastuti as minister of maritime affairs and fisheries to curb illegal fishing, but by 2017, the country had begun to seek help from the UN to fight illegal harvest.[1]

Illegal harvest by countries such as Indonesia speaks to the high global demand for seafood not as a culinary fancy but as a necessary contribution to human nutritional needs. Without some degree of global fishery harvest, that is, many would go hungry. This situation places fishery harvest at the center of a global protein economy. It also demands that recreational anglers be aware that our practice is one of privilege, something we do recreationally, not dependently. This place of privilege, as I have said, demands that we adopt ethical approaches to ocean conservation that account for our positions globally as both stakeholders and stewards in ocean conservation. Part of that ethic, I believe, requires that we understand the implications of designating certain kinds of ocean-dwelling organisms "seafood."

We had been fishing for about forty-eight hours, with another twenty-four ahead of us. Like the other thirty-five or so anglers on board (all men on this trip), I was exhausted, forgoing sleep in favor of more time at the rod. We had each paid about $600 for the trip, and for the most part, none of us had ever met before we had departed on the live-aboard boat from a Florida west coast dock at John's Pass two nights prior, though a few of the regulars were obviously acquainted. Our destination was the Florida Middle Grounds, legendary in Gulf of Mexico fishing lore. The Middle Grounds cover about 460 square miles in the northeastern Gulf, beginning about 80 miles west of Florida, about 120 miles from our port of departure. The Middle Grounds are an array of ancient coral reefs that researchers have dated at more than twenty thousand years old. In the scope of ocean depths, the Middle Grounds are not terribly deep, rising to depths of only fifty feet in some places, while surrounding waters can drop to nearly two thousand feet. The Middle Grounds form giant cliffs and canyons in the Gulf that attract baitfish by the tens of millions and the game fish that feed on them.

Captain Wilson Hubbard started running trips on an

overnight party boat—or sleeper boat—in 1969 from Hubbard's Marina in Madeira Beach, Florida. Captain Jeff Hubbard has explained that the crews of the commercial fishing boats docking at the marina used to boast about the numbers of fish they were catching in the Middle Grounds, so his family started to run a ninety-foot steel-hulled boat out for overnight trips. I took my first trip on the Hubbards' boat in 1991 after visiting their dock one afternoon to see what this infamous Middle Grounds party boat brought to dock. When I saw the mates and deckhands unloading from the holds what had to have been more grouper, snapper, and amberjack—with a few other fish mixed in such as cobia and king mackerel—than I had ever seen from one boat, I knew I had to make the trip. I handed the captain a check for my deposit (in the days when checks were still used for daily transactions) as soon as he came ashore.

He looked at my name on the check and then at me.

"You Milton's boy?" he asked. The question took me by surprise. Milton was my grandfather, and he lived about two hundred miles away in Jacksonville on Florida's east coast.

"No, sir," I replied, "Milton is my grandfather." He paused for a moment, thinking.

"That makes sense," he chuckled. "I lost track of time there. You'd be too young to be one of his boys. I used to fish with him." He put a red *X* in a square on the paper on his clipboard. "You'll be at the stern, but I can't give you a corner. You want a bunk or a seat in the cabin?"

There were about twenty airplane seats anchored inside the main cabin, adjacent to the galley. The bunks were below, near the engine room, but you could lie down in a bunk and only recline in a seat. The bunks were an extra five bucks, but I figured it would be worth it in the larger scope of things.

The position at the stern was an unspoken gift, an

acknowledgment that because of my family name I was part of a confederacy or fraternity of old Florida fishermen. The stern spots were unofficially reserved for the regulars, a group of hard-core fishermen who took the trips to fill coolers with as much fish as possible. These were not the tourists, the fathers and sons on a bonding trip, the college buddies looking to drink beer in the galley and occasionally wet a line so they could boast of fishing prowess. These were not recreational fishermen in the truest sense, nor were they commercial fishermen, though some, I would learn, had commercial licenses. These were meat fishermen; they came to fish. Allowing a tourist or weekend warrior in the stern would have been a flagrant violation of angler protocol (and not the way to treat regular paying customers). The stern was coveted because there was less chance of a one-timer tangling your line; because when the captain anchored, the stern would be positioned closest to the reef line; because bigger fish would often hang back away from the bustle of activity; because you could cast a spinning rod with a plug while the captain was anchoring or pulling anchor and hook a tuna or dolphin before he called for lines in; and because of a lot of other protocols I would never have understood on that first trip. All my life I have wanted to be a regular, whether on one of the Hubbards' boats, on the pier as a kid, or anywhere else in the saltwater fishing world. The captain's gift of that spot is only one of the many reasons I have come to love the Hubbards' trips—the fish, of course, being another.

We fished hard for three days. I slept infrequently, but more than the regulars. I slept down in the hold in my bunk, with the big diesel engines so loud that even with my cassette Walkman turned up all I could hear was the rhythmic pulse of the propulsion system. Nevertheless, when I slept, I slept solidly and comfortably. Moreover, when we fished, I and the regulars, we fished hard and caught fish. Lots of fish.

I brought to boat a nice-sized bonita—also known as little tunny—the most common of the Atlantic tunas, found throughout the Gulf, the Atlantic, the Mediterranean, and the Black Sea. Bonitas are not food fish, though, at least not to many westerners. They are bloody, dark-meated, pelagic fish that many anglers refer to as "boneheads" because they can be voracious and will attack baits in feeding frenzies, often preventing anglers from getting baits past their schools to targeted fish below. They are fun to fight, as are all tuna, but not worth keeping, unless you are going to use the fish as strip bait or marlin bait. I was preparing to release the one I had hooked when one of the regulars hollered for me to put it on his stringer. I asked why since they were not good eating. He responded that he would not eat it, but he could sell it to "a Jap or a Mexican." Not wanting to do anything to reveal myself as an imposter to the back-of-the-boat gang, I had the mate put the fish on his numbered stringer, hoping it would actually be eaten.

To the fishermen working the back of the boat on that trip, and to countless others worldwide, fish are food, nothing more. The old misgivings about the two categories of fish: edibles and trash.

This story could be retold multifold by all anglers, moments when fish not considered food are killed because someone else will eat them—or, more disheartening, when fish are killed because they are regarded as nothing more than trash, as I have described happening on the Skyway Pier and other piers. Or even when they are killed for trophy displays at tournaments. I have also seen paying customers on charter boats and party boats wishing to release a fish and the crew explaining that this boat kills fish, it does not release fish. I have seen this happen with countless species, including marlin, snapper, tuna, various kinds of rays, sharks, and even saltwater catfish. Each incident reminds me that the very concept of "seafood" is as fluid as the ocean itself.

If we are to speak of fish, we now must also speak of seafood, but not as inherently synonymous, as we historically have. There is a familiar understanding among environmentalists, conservationists, patrons of sustainability, and others alert to such things that in English we refer to wild animals as "wildlife" unless they inhabit marine environments, in which case we refer to such animals simply as "seafood." The distinction is important rhetorically as it demarcates the cultural positioning of sea life as predominantly food, neither as wild nor as life.

Recreational anglers must be alert to the power of such language and work to change how we all understand oceanic populations; this is central to a new anglers' ethic, a sport-fishing ethic, a harvest ethic, and ultimately, a global protein ethic. Clark Wolf has pointed out that environmental ethics are forged in an atmosphere of terrestrial ownership, both legal and cultural, often identified as a "land ethic," a way of thinking frequently attributed to Aldo Leopold.[2] Land ethics, he shows, are not simply transferable to marine environments. Relocating concepts such as property lines from two-dimensional surfaces to an endlessly shifting fluid medium loses a bit in translation. However, marine ethics must emerge, ethics that account for recreational and commercial harvest, ethics that are not bound by a culture of seafood but that account for the need and desire to harvest and consume marine animals as food.

Vast quantities of data confirm that the world's fisheries are collapsing. Reductively attributed to overfishing, climate change, and pollution, the rapid decimation of marine species suggests an imminent oceanic singularity. According to the American Fisheries Society, more than 82 fish species in North American marine waters alone are facing extinction; worldwide, approximately 1,851 species of fish—close to 21 percent of all fish species—are at risk of extinction, including more than a third of sharks and rays.

It is one thing to talk about overfishing and the history of fishing pressure—see, for example, magnificent books such as Paul Greenberg's *Four Fish* or Mark Kurlansky's *World without Fish*—but it is another to talk about a larger cultural understanding of what such decimation means and what this moment conveys about the new ocean and the new marine environment. That is, this is not a matter of mere semantic distinction between "seafood" and "wildlife"; this is about larger systems of classification and databasing, about the will to taxonomy. We need to examine the categorical conflict we establish between, say, mammals and fish, between terrestrial and aquatic. We need to understand distinctions between biological extinction and commercial extinction, between survival and thriving.

In the same breath, the very concept of recreational angling will necessarily change in this examination. Recreational anglers are going to have to get used to the idea that fish are going to be smaller, that the reduced populations and the extermination of larger breeding stocks also mean smaller fish. The facts of our current oceanic moment are that there are just not as many big fish out there to catch as there used to be, nor, in many cases, are there as many fish as there used to be. This historic moment cannot be oversimplified into a nostalgia for the good old days of big fish and a bemoaning of how bad the fishing is. This is not merely a call to rectify what has been done wrong. This can be seen only as a tipping point in the culture of angling to redistribute expectation within a new framework. This is the moment to remake what it means to be an angler and to fish. This is the moment for American recreational anglers to embrace changes in ethical positions that will influence angler actions. This is the moment for active cultural change.

Recreational anglers have made such changes in the past.

For example, the emphasis on catch-and-release fishing required active promotion and long-term introduction until the practice was adopted widely and understood as ethically important. More than a hundred years ago, British coarse anglers began using catch-and-release techniques to protect coarse populations in overfished regions, but the practice was localized and specifically contextual. In the United States, catch and release first appeared in the early 1950s as a management practice in Michigan in order to reduce the need for stocking trout waters. The Michigan program evolved at a time when US anglers were popularizing the idea of recreational fishing, of fishing not just to catch, kill, and eat fish, but fishing for the sheer pleasure of the activity—fishing for fun. In this way, recreational fishing and conservation have been bound philosophically, and together they have had a serious effect on how recreational fishing culture has changed over the last century. Of course, early proponents of catch and release, such as legendary fly fisherman Lee Wulff, had been advocating for catch and release since the 1930s, as Wulff's famous line "game fish are too valuable to be caught only once" indicates.

Similarly, recreational anglers have taken up other cultural shifts to promote conservation efforts. For example, anglers now do not think twice about depositing used monofilament fishing line in recycling collection stations instead of tossing old line into the water or into trash receptacles, where it eventually ends up in landfills. Recreational anglers became aware of the environmental hazard fishing line becomes when floating free in the ocean, tangling marine life. We also came to learn that when interred in landfills, monofilament lines break down into microplastics over time, which seep into soil and water and cause a host of environmental problems. Now, recreational anglers expect monofilament line recycling stations to be available at boat ramps, piers, marinas, and even entry points to beaches so that mono lines can be properly recycled. Tossing mono lines into trash cans or overboard—even the smallest trimming when tying a knot—is an angler taboo that is frowned on by the recreational community, but this cultural norm took time to firm up. Now, we see similar efforts in developing conservation ethics and cultural acceptance for recycling plastic lures, as well. Conscientious efforts like these help not only modify angler behavior—which is critical—but contribute to ideological and philosophical modifications. Ultimately, recreational anglers are going to have to embrace a more readily defined and cognizant conservation ethic in which the very idea of seafood and ocean stewardship are redressed within the realities of current ocean conditions. No longer can our nostalgia for a strictly romanticized notion of recreational angling, of fishing for fun, stand independently of a culture of conservation grounded in science. Ensnared in this cultural shift is the power of the word in naming something "seafood" and what that word means locally, nationally, globally.

I confess: I am addicted to McDonald's. I have been since I was a kid. One summer in high school, I worked the late shift at a McDonald's, flipping burgers in the back, taking joy in every rehydrated onion flake I flicked onto those salty sizzling patties. During my breaks, unbeknown to my manager, I'd invent my own McCreations, concocting combinations of patties and condiments in new and extravagant ways (I still believe I created the biggest Mac ever). For several years in high school and college, I worked the morning shift as a yard rat in a large marina, where the marina owners would bring us McMuffins every morning. In the summer of '84 I kept track, and I ate 185 McMuffins between May and September. In college, when I truly had the freedom to do so for the first time,

I probably visited the drive-through six or seven times a week; a friend once counted over two hundred of those beautiful yellow cheeseburger wrappers tossed into the back seat of my '76 Ford Elite. The wrapper archive in the back of my '85 Monte Carlo was even more comprehensive. Lore tells, too, of my mother discovering a cache of several cheeseburgers I had stashed in the Ford's glove compartment—just in case (McPrepper). During my study abroad program in Scotland, I spent a day on a train to get to Glasgow to find those golden arches, and I hung the wrappers of my bounty on my dorm wall like trophies. As a fan of Elzie Crisler Segar, *Thimble Theater*, and Popeye, I always empathize with Wimpy.

Of course, making such confessions today will draw ire from many—and rightfully so. After all, I have to acknowledge that McDonald's is a dietary/environmental/labor nightmare on a GMO bun. I am fully aware of what is in every bite of those burgers. I get that my middle-age spread is due in large part to a lifelong indulgence in those two all-beef patties, special sauce, lettuce, cheese, pickles, onions, on a sesame seed bun (okay, to be fair, it is a result of my indulgences in many things). I know that the "food" is high-fat, high-sugar, high-salt, chemically loaded dreck. I also know that I can get better burgers—with better quality, better taste, better meat—in dozens of other places. I am fully cognizant of what goes into a Chicken McNugget that is not chicken. I am baffled by the logic that a chicken nugget would need to be anything but chicken and that even the chickens are not really chickens anymore, but genetically mutated prenuggets. I also understand the correlation between McDonald's beef consumption and deforestation, intensive farming, and less-than-humane animal harvesting methods. Yes, McDonald's is the world's largest beef consumer, and thereby incriminated in any connection between methane production and global warming. Yet, as an addict, I do not connect the sodium-laden pleasure of those golden fries with the truths I know. Rationality is not a particularity of the addicted.

However, the Filet-O-Fish may change all of this for me. You see, when it comes to hot, greasy, golden goodness, McDonald's Filet-O-Fish is a sizzling treasure, but my oceanic sensibility is likely to overpower my culinary digressions. Let's face it, no matter how masterful one is at frying fish one has caught, one's home-fried filet will never rival the majesty of the F-O-F.

Lou Groen, the McDonald's franchise owner in Ohio who invented the Filet-O-Fish in 1962, may deserve a star on the fast-food walk of fame, but his innovation comes at the cost of detriment to global fisheries. Groen, as the story goes, developed the Filet-O-Fish to provide options for Roman Catholic customers who would not eat meat on Fridays (apparently, the Catholic nutria revolution had not reached Ohio in 1962). Ray Kroc, who joined McDonald's in 1954, saw promise in Groen's idea but had developed his own nonmeat sandwich, the Hula Burger, a grilled pineapple slice with cheese on a bun (really—I'm not making that up). Kroc put both sandwiches on the menu, agreeing that whichever one sold the most would become a permanent part of the McDonald's bill of fare. Of course, the pineapple sandwich was no match for the square of golden-blond fish and tangy tartar sauce on that cloud-fluffy steamed bun, and in 1965, the Filet-O-Fish was added to McDonald's menus around the country. So rather than monoculture pineapple farming, McDonald's opted for trawling.

From the late 1960s until the early 1980s, McDonald's made the Filet-O-Fish primarily from pollock, a name that describes two species of fish in the *Pollachius* genus: *Pollachius pollachius* and *Pollachius virens*, both of which Linnaeus identified in 1758 and had been harvested for food long before. Pollock are a gadiform, or ray-finned fish, which Mark Kurlansky in his brilliant book *Cod* explains: "To the commercial fisherman, there have always been five kinds of gadiform: the Atlantic cod, the haddock, the pollock, the whiting, and the hake." The pollock has historically been considered the lesser food fish of these gadiforms. In the United States, pollock has been used primarily as an ingredient in frozen fish sticks and other frozen seafood products. In 1981, New Zealand McDonald's stores began using red cod in place of pollock, citing better flavor as the reason for the change. The recipe was then shifted world-wide to follow suit. In the early 1990s, New Zealand McDonald's then shifted the recipe to use hoki (a.k.a. blue grenadier) (*Macruronus novaezelandiae*) as a more cost-effective ingredient. At this same time, world fisheries monitors began to identify severe decreases in global pollock stocks.

On September 26, 1996, McDonald's announced the removal of the Filet-O-Fish sandwich from all of its menus. For some of us, this date can be thought of as akin to November 22, 1963, a date for which the devoted ask, "Where were you when you heard?" McDonald's replaced the F-O-F with the new Fish Filet Deluxe as part of its attempt to develop a more enticing menu for a discerning market through the Deluxe line, a branding scheme that failed so miserably that it is considered one of the most expensive product launch failures of all time. As the Deluxe line was failing, consumers were demanding the return of their beloved Filet-O-Fish. I, like thousands of other Filet-O-Fans, signed petitions and wrote letters. On March 22, 1998, McDonald's announced that it would return the Filet-O-Fish to the permanent menu; it readily acknowledged that customer demand had influenced this decision.

By 2007, when the New Zealand hoki fishery began to decline, McDonald's adjusted to a predominantly Alaskan pollock–based recipe for the F-O-F, citing declining Atlantic cod fisheries as influencing its decision to not use cod as the primary ingredient. According to its web pages, McDonald's works to adhere to Marine Stewardship Council (MSC) recommendations regarding sustainable fisheries. The MSC is an international nonprofit organization that addresses "the problem of unsustainable fishing, and safeguarding seafood supplies for the future." It is one of the world's leading supporters of catch-share programs. While the MSC reported in 2013 that McDonald's suppliers adhered to sustainable practices, recent calls by environmental groups to curb commercial harvesting in the Bering Sea canyons—where much of McDonald's F-O-F pollock is taken—call into question the viability of continued harvesting in the region. Organizations such as Greenpeace and the North Pacific Fishery Management Council, which has also implemented catch-share programs throughout the Northeast, have been vocal in questioning the long-term risks of continued harvesting in the area. Many groups have been examining the need for protecting the Bering Sea canyons.

Alaska senator Lisa Murkowski, who is cochair of the Senate Oceans Caucus and who often speaks about Alaska's sustainable fisheries and fisheries management, wrote to McDonald's president and CEO Donald Thompson imploring him to not take seriously the concerns Greenpeace had voiced about the sustainability of the Bering Sea canyons fishery. Murkowski specifically explained that when it came to the ways in which Greenpeace and other environmental groups had characterized the fishery, "nothing could be further from the truth." She elaborated: "There are two entities with responsibility for federal fisheries management off the coast of Alaska: the National Oceanic and Atmospheric Administration (NOAA) and the North Pacific Fisheries Management

Council. The North Pacific Council is recognized through the country for its dedication to utilizing scientific research and data when making decisions. . . . Both [agencies] have recently concluded this portion of the Bering Sea is not at risk from fishing activities." Likewise, in the same press release in which she shared her letter to Thompson, Murkowski included a screen shot from a McDonald's web page that declared, "The Bering Sea is one of the best places to catch wild Alaskan Pollock." On her Facebook page on July 31, 2014, Murkowski said, "McDonald's own website brags about our successful, sustainable Bering Sea fishery—so why would they reconsider based on pressure from environmental groups whose claims have no basis in fact or science?" I'm not sure how Murkowski makes the correlation between the "best places to catch wild Alaskan Pollock" and the boasts about sustainability; her claim seems contrived.

Greenpeace responded by claiming that Senator Murkowski was misrepresenting its efforts. On August 13, 2014, Greenpeace blogger Jackie Dragon wrote: "Our campaign is not calling for a boycott of Alaska pollock, nor is it an 'anti-fishing campaign.' This is about a place, and not about any individual fishery. It is about the long overdue need to protect a portion of the Green Belt—the crowned jewel of the Bering Sea where the world's largest underwater canyons harbor the lion's share of coral and sponge habitat, essential fish habitat that is important for numerous commercially important species, and is a vital component of a healthy, functioning ecosystem."

It seems to me, one of the Filet-O-Faithful, that Senator Murkowski's rather broad-ranging claims about sustainability actually fail to account not just for the longevity of the Bering Sea canyons fishery, but for the ability of McDonald's to sustainably provide customers with one of their favorite products. That is, despite Senator Murkowski's overoptimistic claims about the failure of environmental organizations to engage the science that shows successful sustainability, there are in fact specific scientific data showing that the Bering Sea canyons are a vital ecosystem for many other organisms beyond those harvested for commercial purposes. Research from organizations such as the Marine Conservation Institute shows that impacts from commercial harvesting technologies such as trawling have ecological effects beyond the targeted species. Two of the largest canyons and primary commercial fishing zones in the region, for example, the Zhemchug and the Pribilof, are adjacent to an unprotected region known as the "Green Belt." This region of the Bering Sea, where the continental shelf drops into the North Aleutian Basin, is a critical site for primary production, a foundational process in large global systems. In the Green Belt, the flow of colder Aleutian waters mixes with the shallower water of the continental shelf to provide ideal conditions for perpetual production of phytoplankton, and thereby photosynthesis.

Sylvia Earle, renowned marine biologist and former scientist -in-residence at the National Geographic Society (and author of the landmark books *Sea Change: A Message of the Oceans* and *The World Is Blue*), has named the Green Belt region a "Hope Spot" in conjunction with Mission Blue, an organization Earle formed in collaboration with the TED organization and her 2009 TED Prize. In her recently released documentary *Mission Blue*, Earle explains that "Hope Spots" are large areas of ocean that need protection, asking that they be designated as no-fishing and no-dumping sites. Earle's call for the Green Belt to be protected in this way comes from years of research—the very science that Murkowski claims those wishing to protect the area are not using.

Research by the Marine Conservation Institute also shows

that commercial harvesting is responsible for the removal of more than a million pounds of deepwater corals and sponges each year. Other research-based organizations, such as NOAA and the Alaska Fisheries Science Center, have confirmed that the Green Belt contains the richest habitat of deepwater corals and sponges. Because of the abundant corals, sponges, and phytoplankton, this region is a primary breeding, foraging, and nursery ground for the very species that Murkowski encourages McDonald's to continue harvesting. The logic seems baffling to me in terms of both ecological and economic sustainability. If the foundational elements of the system that drives and sustains McDonald's and other businesses (not to mention Alaska's own economic gain) are not protected and in turn risk destruction by the methods used to harvest the economically viable portion of the system, then the harvestable resource inevitably becomes a temporary and limited resource. I suppose this is part of Earle's overall point: the need to consider entire systems, not merely the parts that conveniently serve a specific interest. Thus, recreational anglers need to enfold other components of ocean conservation into the new ethic recognizing the very ecologies bound up in the ocean. Coral, for example is crucial for healthy fisheries and healthy oceans, as are menhaden.

Likewise—and again despite Senator Murkowski's grand claims of her state's sustainability achievements—if we look at the Marine Conservation Institute's *SeaStates 2014* report, Alaska ranks at the bottom among all US coastal states in terms of coastal protection. Alaska does not maintain a single no-take area in its waters. NOAA, the very organization producing the science on which Murkowski relies, has specifically indicated that the Green Belt stands to suffer the greatest reduction of any area in the Bering Sea because of commercial fishing technologies. NOAA predicts that if present fishing rates continue, Pribilof Canyon alone is at risk of a 50–75 percent reduction in living structures such as corals. Also, again despite Murkowski's claims, the Green Belt is the only region within the Bering Sea with no applied protective policy or practice.

Of course, the slippery slope in this argument is that closing the Green Belt to commercial harvesting will inevitably promote closure mentality and increase the fishing pressure on another fishery. Similarly, it is likely that the commercial fishing industry will find some way to make up for any decline in product retrieval that a ban would create. Of course, to say so inevitably also opens the door to the potential that such a closure could embolden governmental agencies to close off other fisheries as well, again supporting the very idea of closure mentality. This is a complex debate regarding the rights to harvest and the processes—both research based and cultural— employed to determine where such closures might be implemented. Debates such as the current contested harvest rights in North Carolina's battle over inshore netting will inevitably force us to more thoroughly and more systematically address the structures that determine fisheries policy.

Like Senator Murkowski, I am confident in science and science-based policy, but I also place value in common sense, and it seems to me that common sense shows that Senator Murkowski's science conveniently crops out a substantial part of the picture. There is a bigger context here beyond the sustainability of one commercial fishery, and it is not just about a sandwich.

Ultimately, recreational anglers will need to address the discrepancies between recreational harvest policies and commercial harvest policies, including not only the economic imbalance of catch shares and the immediate consideration of overharvest, but the global illegal, unreported, and unregulated harvest. Recreational anglers will need to find a voice in concerns about global protein economies and the role of aquaculture and fish farming as sustainable providers of marine protein for those in need. Recreational anglers, that is, will need to take up all matters of global ocean conservation for the simple reason that any engagement with the world's ocean will affect the culture of recreational angling.

In many ways, it would be convenient, if not prudent, to rewrite my concerns here for recreational angling and ocean conservation strictly in the wheelhouse of global commercial harvest. Certainly, such discussion will continue to be necessary, but it becomes valuable only as it emerges from an evolving recreational angling ethic that has yet to galvanize. Toward the end of such an ethic, then, I set aside the turmoil of commercial harvest for the time being to return to the culture of recreational saltwater fishing, to the satiety of the saltwater life, to the deep blue of thought.

Globally, there can be no doubt that fisheries harvest will always be bound to a protein economy to some degree, whether the harvest be recreational, commercial, industrial, artisanal, or other. That is, humans will continue to harvest fish for food. Because of this inevitable association and naturalized use-value, recreational anglers need to begin to address the distinctions that set ocean harvest apart from other harvest mechanisms in the total global protein economy. For example, we can easily point to the fact that terrestrial contributions to the global protein economy are fulfilled primarily through farming practices. We know, for example, that relying on wild harvest of animals such as deer or turkey would never provide enough protein to fulfill demands. Likewise, as we know from historical evidence, such an attempt to provide that degree of protein globally through wild harvest would quickly render the target species extinct. Consider, for example, the overhunting of bison in the United States for commercial purposes, primarily for hides, and the relatively rapid decimation of that population. Bison were sustainably hunted by Native American populations, but the introduction of commercial harvest methods spurred by economic incentive left populations nearly extinct by the 1880s. Bison, like deer and other terrestrial herd animals, could not be harvested in a way that would efficiently supply the protein economy. Likewise, humans did not find harvest limits that would sustain the populations of those wild herds, particularly within a cultural mind-set that assumed the vast herds were inherently unlimited. Thus, we were forced to domesticate. And while farming operations in the United States are rarely ecologically ideal, they do support the sustainability of wild game populations by deferring protein fulfillment to farmed animals as the primary protein source. Debates about the authentic "animal-ness" of farm-raised organisms versus wild animals will need to continue to unfold, but there can be little question that farmed animal protein is necessary in the global protein economy. Similarly, from a strictly terrestrial standpoint, the primary contribution to the sustainability of many wild game species was the shift to domestication and farm raising.

Look, it is this simple: if we are to continue to include fish as a primary contribution to the global protein economy while simultaneously investing in ocean conservation and sustainability, then we need to promote aquaculture and mariculture in parity with terrestrial farming. This is no simple task. The demand for fish protein cannot be sustained by wild harvest, even under the current restrictions placed on commercial harvest in the United States and other countries. I recognize that many will see such a claim as an attack on commercial fishing and the livelihoods of those who depend on commercial harvest to make a living; however, my argument is not against the idea or practice of commercial harvest, but against the current focus on wild harvest rather than farmed harvest. That is, I seek a shift in the baseline regarding (1) where we harvest, and (2) our cultural understanding of where food fish come from. Farmed instead of wild for food.

Fish farming in its current iterations is far from the ideal answer to the global need for fish protein. Current practices cannot meet the scale of need for fish protein. As Jared Diamond notes in *Collapse: How Societies Choose to Fail or Succeed*, demand for seafood continues to increase globally, notably doubling in China over the last ten years.[3] According to Diamond, fish account for 40 percent of all protein—plant and animal— consumed in the third world. Fish, for example, are the primary protein source for more than a billion Asians.

Given the massive need for fish protein, fish farming will necessarily need to become more commonplace and certainly more ecologically feasible. The Food and Agriculture Organization of the United Nations (FAO) puts forward one of the most commonly cited definitions of aquaculture, explaining that it

> is understood to mean the farming of aquatic organisms including fish, mollusks, crustaceans and aquatic plants. Farming implies some form of intervention in the rearing process to enhance production, such as regular stocking, feeding, protection from predators, etc. Farming also implies individual or corporate ownership of the stock being cultivated. For statistical purposes, aquatic organisms which are harvested by an individual or corporate body which has owned them throughout their rearing

period contribute to aquaculture, while aquatic organisms which are exploitable by the public as a common property resource, with or without appropriate licenses, are the harvest of fisheries.[4]

"Aquaculture," then, can be understood as an umbrella term that encompasses many methods and approaches. Mariculture, for example, is aquaculture within maritime environments, often in open-ocean environments or controlled estuarine environments such as marine ponds.

Aquaculture is by no means a new concept. Evidence shows that various cultures manipulated local environments to raise fish populations as early as 6000 BCE. In the United States, farming of oysters and trout had become common in some regions as early as the mid-1800s. Open-ocean aquaculture is one of the most recent developments in mariculture. Even now, though, we employ only a few methods of mariculture, and most technologies are in their infancy and will require substantial evolution before becoming feasible farming approaches. In 2010 the Congressional Research Service (CRS) published a report titled *Open Ocean Aquaculture* that defined these open-ocean farming operations:

> Open ocean aquaculture is broadly defined as the rearing of marine organisms in exposed areas beyond significant coastal influence. Open ocean aquaculture employs less control over organisms and the surrounding environment than do inshore and land-based aquaculture, which are often undertaken in enclosures, such as ponds.[5]

Few of these farms exist, as they require substantial energy and capital to maintain; yet many see them as holding significant possibility for the global seafood market and protein economy. They are, however, so new in concept that regulatory issues have also limited their development. The CRS report explains:

> When aquaculture operations are located beyond coastal state jurisdiction, within the U.S. Exclusive Economic Zone (EEZ; generally 3 to 200 nautical miles from shore), they are regulated primarily by federal agencies. Thus far, only a few aquaculture research facilities have operated in the U.S. EEZ. To date, all commercial aquaculture facilities have been cited in nearshore waters under state or territorial jurisdiction.[6]

What is meant, though, by "federal agencies" is the eight fishery management councils. For example, despite some opposition from environmental organizations and some commercial fishing organizations, in 2009 the Gulf of Mexico Fishery Management Council—the same group embroiled in issues of poor red snapper management—voted to approve open-ocean aquaculture permits. From a strictly recreational angling perspective, the control of policy by management councils in all three facets of the harvest puzzle—commercial and industrial, recreational, and farming—seems particularly worrisome.

Like all other components of fisheries regulations, mariculture has faced significant difficulties in federal policy making. In 2009 Nick J. Rahall (D-WV) introduced H.R. 3534, the Consolidated Land, Energy, and Aquatic Resources Act of 2010, to the House of Representatives. H.R. 3534 instituted policy for the development of renewable energy resources on public lands. Section 704 of the bill proposed rescinding the right of the fishery management councils to oversee offshore aquaculture policy and permitting. When the bill passed the House in 2010 with a vote of 209–193, it did so without the inclusion of section 704, and the fishery management councils retained oversight of offshore aquaculture. Six months prior to the House vote, though, Representative Lois Capps (D-CA) introduced H.R. 4363, the National Sustainable Offshore Aquaculture Act of 2009, which directed the secretary of commerce to "establish an Office of Sustainable Offshore Aquaculture in the National Marine Fisheries Service at National Oceanic and Atmospheric Administration (NOAA) headquarters and at satellite offices in each of NOAA's regional fisheries offices."[7] In 2010, Senator David Vitter (R-LA) introduced Senate Bill 3417, the Research in Aquaculture Opportunity and Responsibility Act of 2010, which "prohibits any executive agency or any Regional Fishery Management Council from developing or approving a rule, regulation, or fishery management plan to permit or regulate offshore aquaculture until the date that is three years after the date of the submission of the reports required by this Act."[8] The bill called for extensive studies of the impact of offshore aquaculture before permits might be issued. Since its introduction to the House on May 25, 2010, the bill has been read twice and referred to the Senate Committee on Commerce, Science, and Transportation, but no other action has been taken as of this writing. All of this leaves the future of offshore aquaculture in US waters uncertain.

Nevertheless, it is not just the legislative mess that strands mariculture in a sea of uncertainty. Despite the history of the idea of fish farming and the limited implementation of that idea into practice, fish farming technologies have a long way to go before they become a feasible answer to the global protein economy's need for fish protein. Scalability will always be a hurdle. Likewise, many current methods are as harmful, if not more so, to local environments as industrial harvest of wild populations. Given the relative infancy of offshore aquaculture practices as well, there are few data to confirm ecological and economic feasibility. Simply put, we just don't know what might happen with large-scale offshore farming. Certainly, we can intuit some problems that will need to be addressed, and many environmental groups that oppose farming are quick to point to a few of the more evident assumptions as assumed risks. For example, current variants of mariculture face several primary ecological and economic difficulties. Many opposed to offshore aquaculture, for example, assume mariculture will result in some of the same ecological problems that inshore and onshore aquaculture efforts have faced, including the polluting effects of fish farming. Inshore farms, for example, have been criticized for the spillage of nutrients and biowaste (fish shit) into local environments in quantities that exceed natural levels, which in turn affect benthic layers of marine environments. Questions remain as to whether offshore farms will release more biowaste and whether that waste will be more readily absorbed by open-ocean water. The truth is that we just do not know; however, not knowing is not a reason not to do. Research, experimentation, and discovery are crucial at this stage.

Similarly, some skeptics voice concern not only about the waste output of mariculture, but about the input of materials into the system as well. Offshore farming opponents express concerns about the introduction of antibiotics, steroids, and other pharmaceuticals to farm environments and the results of those chemicals dissolving into open-ocean waters. Along these same lines, some worry about disease conveyance within a given farm and between farmed animals and local wild populations.

Likewise, oppositional positions question the effects of farmed fish escaping the confines of farm encasements. How will farm-raised fish interact with wild populations and affect their genetic makeup? In a 2004 article in *Wired* magazine,

Charles C. Mann explains that the outer netting of mariculture cages is "made of Spectra, a superstrong polyethylene fiber used by NASA to tether spacewalking astronauts to the mothership. Wrapped tautly around the frame, it leaves no slack for predators to grip, but the material is built to withstand the most ferocious attacks."[9] Predatory attacks are not the only concern regarding damage to the cages that might lead to escape. Open-ocean conditions can produce extreme storms, currents, and high-energy conditions that might damage or destroy mariculture cages. Of course, too, there is no way for a mesh cage to prevent fertilized eggs from floating free of the enclosure. Ultimately, there can be no way to guarantee that fish will not escape their pens.

Perhaps most significant, though, is the question of energy distribution. As offshore farming is likely to produce carnivorous, and often predatory, species for a protein market—salmon, tuna, cobia, mutton snapper, halibut, haddock, cod, and more—the ratio of energy input to energy output is problematic. Carnivorous fish require substantially more energy input than they produce as output. For example, a salmon requires several more pounds of prey fish for energy than the salmon itself weighs. Energy decreases as it moves through trophic levels; therefore, fish occupying higher trophic levels require massive amounts of energy intake in order to survive and grow. This translates into an unsustainable formula wherein aquaculture requires more energy—usually in the form of prey species—going into the system than comes out through the raised fish. This is both cost prohibitive and ecologically dangerous. Solutions to energy depletion will stand at the fore of developmental needs for mariculture.

These are technological problems, questions to which we do not yet have answers, but problems that can be solved.

Because they are technological problems, they are likely to be easier to solve than the cultural questions that will inevitably accompany mariculture as a solution to harvest issues in light of global protein economy needs. Recreational anglers will need to advocate for farm-raised fish as not just a viable answer to global protein needs but as a key component in long-term ocean and fishery population sustainability. Adopting a farm-raised-for-food ethic includes the need for a cultural shift in understanding the individual wild animal as opposed to a farm-raised food animal. This also requires accountability for another inconvenient science if we intend to protect global fisheries by way of growing food fish. We cannot accept conservation science on one hand and on the other dismiss science that improves aquaculture and the efficiency of food growth, including genetic modification. In terms of the global protein economy, cultivated fish populations and even genetic modification of those populations will likely be a necessary part of aquacultural approaches to providing sufficient protein. These, however, are debates that will need to unfold sensibly.

Ultimately, we need to come to terms with the difficult position that in order to protect recreational saltwater angling, the fish populations and ocean environments on which recreational angling depends, and an ethical culture of harvest, we must concede fish populations as necessary in the global protein economy. In such an acknowledgment, we must also develop an ethic that promotes methods such as mariculture as an alternative to wild harvest while simultaneously finding a balance for management of wild harvest that accounts for recreational needs and a reduced and monitored commercial harvest. All such ethics will inevitably be bound up in complex management policies, as well as deep-seated philosophies of what it means to be a recreational saltwater angler.

7

Wayfaring and the Age of Binge

Wayfaring, I believe, is the most fundamental mode by which living beings, both human and non-human, inhabit the earth.

Tim Ingold, Lines: A History

For each of several months last year, I received an email survey from the Florida Fish and Wildlife Conservation Commission (FWC) asking if I had fished in saltwater in the month since the last survey. The survey asked how many times I had fished, where I had fished, and what kinds of transportation and equipment I had used. If I responded that I had fished during that time from a boat, the survey prompted a series of questions about individual fishing trips, asking where I departed from, how many different spots I targeted, travel time and distance between spots, and so on. The survey was designed to plot points on a chart, to consider those points substantial for one reason or another. Perhaps the researchers who designed the survey sought to identify which ramps and facilities anglers used most frequently, or which waters anglers fished most heavily. Perhaps they looked for regional patterns regarding access and usage, heavy-traffic areas as opposed to light-use areas. I suppose, too, the surveys could help estimate how much money I was spending on these fishing trips in terms of things like fuel costs for trailering a boat or running a boat between fishing destinations. The survey gathered data, quantified my time, my expense, my territorial range, reducing my experience to a point on a chart lost among thousands of other points and tables filled with numbers.

I responded faithfully to each survey, noting that research and data are important, that such data contribute to what we know, and to what we need to know about recreational angling practices in the state that claims the title "Sport Fishing Capital of the World." Data and science, I reminded myself with each survey question answered, are critical to maintaining this way of life, to improving the ocean conditions where we fish, to the ethical obligation of conservation. Without science, there is no conservation. Dutifully, I checked boxes and clicked points on maps, reducing hours on the water to mensurable information. The act was clinical, devoid of any value for that time on the water, the experiences on the water inconceivable and irrelevant to the data. The sublimity, beauty, and spirituality of the water and the fish erased by the movement from point A to point B, plotted on a computer screen with a simple survey click. My adventures pixelated.

For a significant population of saltwater anglers, this is how we think about fishing: a series of destinations, points to reach. We drive hurriedly while choking down gas station coffee and fast-food biscuits and honey buns, trying to get to the ramp, dock, or marina as quickly as possible. Home to ramp. The drive is an annoyance between a starting point and the destination, a consumption of time that intrudes on time fishing (weren't we promised teleporters by now?). Stops at the gas station and bait shop are necessary but are detractors just the same. Other anglers in line to pay for bait and beer are unspoken competitors. They might beat us to the ramp, get to our spot ahead of us, or just have a few more seconds on the water than we do. Hurry! Get to the ramp before they do! We start our motors, jam the throttle down to get to the *X* on the chart as quickly (or as fuel efficiently) as possible in order to spend as much time as possible at the destination. Once we reach the destination, the objective is not to fish but to catch fish. If we don't fulfill the objective in a reasonable amount of time (a fluctuating variable), the motor is reengaged and the next *X* is targeted. For most saltwater anglers, this is standard operating procedure, not necessarily by choice, but by constraint of time and a culture of urgency. You see, most who fish in saltwater do so limited by time; they may do so only once or twice a year, or maybe, if they are lucky, once or twice a month. Thus, our culture of urgency drives us to maximize the time we have by rushing from point A to point B. In our current cultural outlook, knowing where to go always trumps discovering where to go. The constraint of time compounds this sense of expedience, as well, because we do not have the luxury of selecting the day or the time to depart toward point B. We go at the time we have designated as the time to go. We go during vacation, on the weekend, or whenever we can rationalize that this is the time to go. Moreover, we fish with the fervor and haste of trying to achieve all our angling desires and fantasies fueled by media representations in that exact moment; otherwise the trip has been a failure.

I think of this kind of fishing as investment fishing or binge fishing, a mentality born in a culture of cost-and-return thinking and exacerbated in a culture of binge. This is the way of thinking that drives everyday activities, rushing from point A to point B. We hurry from home to work and back home, trying to accomplish as much as we can in each setting. We establish vacation plans this way: how quickly can we get from home to the water; what's the shortest route, the least intrusive route? It is the mentality that allows us to disregard the adage "it's not the destination, but the journey that counts."

In the saltwater sportfishing industry—as in any industry—this is the very mentality that drives innovation. It leads to technological advancement in boat design, tackle development, clothing design, and so on. Innovation, in this sense, emerges because the industry has to answer the urgency that our culture demands. Our product will get you there faster; our product will catch more fish; our product will improve your chances of success in your fish quest; our product will make others see you as a true angler. Angler identity is bound up as much in the culture of the angling industry as in the act of fishing; what you have and wear are as much an indication of angling knowledge as experience on the water. The urgency mentality, of course, also drives consumption, economic and ecological. Without the investment and urgency mind-set, we would likely not progress as an angling community—if what we mean by "progress" is to have more anglers fishing with a sense of urgency. Progress seems to suggest increase rather than reduction. The American Sportfishing Association, for example, promotes campaigns to increase the numbers of registered anglers nationwide. Grow the culture and grow the industry in order to preserve the culture and industry. Moreover, while some who would cast innovation and technology as the bane of ecological preservation might argue that curbing innovation might be a good thing, such nostalgic thinking fails to account for an encompassing view of technology and fails to acknowledge the reality of the present. Such thinking denies the value—economic and cultural—of angling innovation and a growing population of anglers. We will not, should not, embrace a nostalgic view of fishing as an objective for the angling community. The past is not the future of recreational saltwater fishing. The future of the world's ocean lies not in the past. Reclamation thinking risks unrealistic faith in nostalgia. The future of the world's ocean lies not in point A or B, but in the ways in which we inhabit the spaces between A and B, not in looking ahead or behind, but in looking around.

The saltwater angler's most powerful method of contributing to the future of the world's ocean lies not only in numbers and capital, not only in industry and politics (though from an advocacy position, those certainly do not hurt the agenda). The saltwater angler's contribution lies in a fundamental respect for the ocean and in our actions associated with all things oceanic. The saltwater angler's ethical contribution to the future of the world's ocean requires nothing more than critical observation. This insight derives from habitation, from inhabiting a place rather than visiting it, from living in the spaces between point A and point B. As Tim Ingold explains, "by habitation I do not mean taking one's place in a world that has been prepared in advance for the populations that arrive to reside there. The inhabitant is rather one who participates from within in the very process of the world's continual coming into being and who, in laying a trail of life, contributes to its weave and texture."[1] Saltwater anglers, I propose, live not as tourists but as wayfarers, not as visitors to the ocean but as residents in its spaces. Even when not physically in its presence, the saltwater angler lives in the oceanic aura, its presence guiding our thinking about how we walk through the world. The saltwater angler must be invested in all aspects of recreational angling, from the ephemeral moment of the fish's strike to the absolution of the release; from the emancipation of land-based thinking to a more fluid understanding of our relationships with the ocean and, in turn, the world. We find, as anglers, a way to incite a contemplative, wandering perspective amid a seemingly entropic global condition. Saltwater anglers, I mean to say, adopt anglers' ethics not only as an investment in the future of the world's ocean, but as a teleological acknowledgment of our inextricable bond to that ocean. The saltwater angler, then, no matter his or her background, lifestyle, beliefs, or commitments, lives—at least in part—as wayfarer.

You see, the FWC survey, in its data collection, reveals that terrestrial linearity binds our thinking about ocean and fishing. Movement from point A to point B, we assume, occurs across a flat surface, as on land. My path from the boat ramp to a waypoint on my chart is mapped the same way as the path from my home to the boat ramp, a movement over a relatively stable surface. We internalize this movement in how we then think of oceanic encounter, casting the marine environment not as fluid space, but as terrestrial solidity across which we move. To stay at the surface in our thinking, I believe, limits how we understand the ocean. To see only the path from A to B abridges our ability to perceive where that trajectory of navigation fits in a larger oceanic world. Perhaps the cultural A-to-B navigation mind-set protects us from the oceanic sublimity, from the overwhelming endurance needed to embrace such incomprehensible vastness, lets us navigate with a narrow field of vision, with blinders. Perhaps it shows us the painless path without the mental work of understanding just how astronomical our oceanic lives are.

The FWC survey is not alone in thinking about the movements of anglers. In 2018 the journal *Marine Policy* published research by Edward V. Camp, Robert N. M. Ahrens, Chelsey Crandall, and Kai Lorenzen in an article titled "Angler Travel Distances: Implications for Spatial Approaches to Marine Recreational Fisheries Governance."[2] The article makes the cutting-edge argument that socioeconomic data now influence recreational fisheries management in ways that need to account for place-based considerations. This research shows how recreational fisheries management has historically used spatial information regarding movement of fish species to establish

management policy—for example, distinctions between migratory populations and resident populations, as well as seasonal
population distinctions—but only rarely has such policy been
influenced by data regarding the movements of anglers.

To facilitate the use of such data, Camp and his colleagues
assess information regarding angler travel in six popular
fishing areas in Florida. Their research examines angler travel
distances—where anglers come from to fish the observed
waters—and compares the travel with corresponding distinctions between targeted species, destinations, and years. This
fascinating research shows us that anglers who target snook,
for example, tend to travel less than thirty kilometers, while
those targeting red snapper may come from more than two
hundred kilometers away to fish. These kinds of data, in and
of themselves, are eye opening; however, the real benefit is the
methodology Camp and his colleagues have developed, allowing researchers from specific locations—say, a specific county
in Florida—to map the travel and range of anglers who fish
the waters of that county. This research is valuable in so many
ways, from considerations of socioeconomic angler movement,
to tourism planning, to harvest limits. For example, such data
might inform decisions regarding possible resident limits and
seasons versus out-of-state angler limits and seasons, as well as
other, similar models.

I wonder, too, about how this kind of research might contribute to our understanding of angling and binge or urgency.
Imagine, for example, a county that is shown by way of place-
based data collection to service more anglers from out of state,
or more out-of-county anglers than local anglers. The same
data might show that these anglers, in order to target a specific
species of fish, are more likely to trailer their own boats than
employ local for-hire fleets. Hypothetically, such information

might influence decisions regarding development of more efficient marina and ramp facilities designed for higher capacity and better efficiency in water access and time at the dock. Such approaches might appear to answer demands of angler pressures at such facilities, but they might subsequently also promote angling urgency. Similarly, the same research might produce tiered harvest structures to encourage out-of-state anglers to fish in the area because of increased bag limits over those of local anglers, or vice versa.

As with any data, the critical question will be how policy makers use the data. In addition, we need to have these data in order to make informed policy decisions. Likewise, we cannot address the data about recreational angling without engaging in a conversation with the culture of recreational fishing. Ultimately, the data and the culture will need to influence one another in a tenuous conversation until they can feasibly function together.

Camp and his coauthors are right to point out the complexity of the systems in which anglers and angler data circulate and the necessity of incorporating place-based thinking in how we consider those systems. The study maintains an understanding of angler movement as being between point A (the angler's origin point) and point B (the angler's destination). However, the study also expands the linearity of such thinking to recognize that these paths intersect with many other variables within complex systems, including time limits, seasonal restrictions, fishery regulations, fishery populations, fish movement from undefined points A, B, and C, and so on, not to mention as yet unknown or discordant factors in the system.

Ingold explains that measuring how fast one moves between point A and point B is not relevant to the wayfarer. He points out that "what matters is not how fast one moves, in terms of the ratio of distance to elapsed time, but that this movement should be in phase with, or attuned to, the movements of other phenomena" of the world.[3] To ask "how long does it take" to move from point A to point B is relevant only if the destination is already known and the traveler desires, according to Ingold, only "to spend his time *in* places, not *between* them. While in transit, he has nothing to do."[4] The wayfarer, I argue, is always already in the place that matters, even when that place is miles from the water, but the water occupies and guides the wayfarer's thoughts.

Often, when the thirst to feel better connected to the ocean takes hold of me, I will push the boat offshore, out of sight of land, piloting with no destination in mind. GPS turned off, charts stowed. Just moving. I fantasize doing so until I run out of gas, erasing the possibility of return, but I can never bring myself to do so. After some time, I shut down the motors and drift, with only a vague notion of where I am, of where the boat is sliding. As I wend about, I fish, I relax, and often I nap. To wake at sea unsure of one's precise location can spark mental stimulation—particularly depending on the condition of one's awakening. As often as I have woken comfortably warm in the sun and calm seas, I have been jostled awake by choppy waters, rain, and wind. I have been startled awake by unfamiliar sounds, and once by a solid smack on my hull followed by a tremendous splash from the same side of the boat, leaving me wondering what creature—a turtle, shark, manatee, seal, or mermaid—my drifting hull, another scrap of debris afloat on the water, had spooked. To wake in the dark at sea can be quickening, heart stirring, a testimonial of one's smallness—particularly if one is also in a small boat. The experience, of course, is much different when one chooses to be in the ocean in this way than if one finds himself or herself there by accident of misfortune. Think, for example, about the experiences Stephen Crane describes in

"The Open Boat" of the four men in the small lifeboat following the sinking of the steamer *Commodore* as opposed to the experiences Hemingway gives us of the small boat in *The Old Man and the Sea*. Of course, because I intentionally choose to create this experience, it can never be truly authentic, never anything more than a specific kind of experience, one rendered safe by the context I create. Much like a roller-coaster ride, which always maintains a foundation of safety; no matter how hard we imagine the cars flying from the track, we know they will not. We are confident that the experience of the ride will always be safe within reasonable expectations. Certainly, too, such a choice is distinctly different from the experience of those shipwrecked at sea, with no predetermined safety valve. Hence my inability to run the gas tank dry and commit to the ocean.

Indeed, our literary history has provided a wealth of second-hand and fictional accounts of what it might be like to wander aimlessly on the ocean, but these accounts mediate our experiences in limiting ways, just as early tales of sea monsters and vast whirlpools contributed to a cultural fear of the ocean. For the recreational angler, wayfaring—as both a guiding philosophy and physical practice—can help us become acquainted with the ocean and certainly, in my experience, can help us become better anglers, providing firsthand, intimate understanding of the ocean.

Wayfaring, Ingold shows us, is distinct from preplanned navigation. The wayfarer is not concerned about departure and arrival, about destination, but about living in the moment of travel. The wayfarer inhabits the saltwater fishing life and the saltwater fishing life is that of the wayfarer.

As Ingold puts it, "The wayfarer is continually on the move. More strictly, he is his movement."[5] For the wayfaring angler, the movement from point A to point B is not merely a transitional activity but a way of being part of the fishing life. The movement between points is not the defining characteristic of the travel or the traveler, but part of a larger anglers' philosophy and ethic. If recreational angling is truly the most contemplative of pastimes, then we must embrace the idea that such contemplation occurs not only when waiting for the bite, but when moving in and between places.

Yes, I recognize that this might sound as if I am preaching a kind of at-one-with-the-world, hippie neo-spirituality, but that is not my intent. In addition, I see that my line earlier that the wayfarer inhabits the fishing life and the fishing life inhabits the wayfarer is tantamount to Chirrut Îmwe's "I am one with the Force and the Force is with me" mantra. However, I am not interested in mantras or philosophies (though if I were, I would probably lean more toward the Sith than the Jedi, if truth be told; the self-righteousness of the Jedi always pisses me off); rather, I extol wayfaring as a mind-set and method in order to emphasize a more comprehensive awareness of the spaces in which we fish. That is, wayfaring allows us to be a part of the places in which we fish rather than only ever being visitors, tourists, or aliens. Wayfaring teaches us that our presence in the places where we fish is relevant to the protection of those places, and that we must take up an ethic of protecting those places as well as those through which we travel to get to where we fish. To protect such places is simultaneously to protect ourselves, our way of life, our community, our future, and our angling heritage. This also means taking the responsibility to know these places intimately beyond just their fish. As such, anglers learn patterns and rhythms, seasons and migrations. Anglers learn the ecologies and inhabitants of these places, learn to understand where the fish-we-pursue and we fit. Knowing where we fit leads to knowing how to be part of a place rather than a visitor who extracts from that place. Certainly, there will always be places we visit, but there must always be places we inhabit, places we know as our home waters. The importance of each is crucial; we must remember that when we visit, we always visit where someone else inhabits.

As a method, wayfaring promotes observation, a more scientific accounting of the waters in which we fish. Wayfaring does not imply arbitrary. In fact, the word itself means to fare along a way, to travel along a route. We fare well—or farewell—along the way, marking waypoints as more than simply spots along the way, but as places themselves worth our attention. Wayfaring, then, is to note the points and details along the way, to observe the parts that make up the whole, to account for the details. Only through this kind of participatory engagement with the ways we travel and fish will we be able to develop, maintain, and encourage ocean sustainability.

Ingold makes the important point that "the wayfarer has to sustain himself, both perceptually and materially, through an active engagement with the country that opens up along his path."[6] For the angler, the ocean, and the fish, this kind of sustainability is critical. At the apex, the angler's concern for ocean sustainability must be paramount, as must the implication that such sustainability results from and supports the sustainability of the fish we pursue and the ecologies in which those fish participate. Likewise, the sustainability of the marine environment and of the organisms that directly participate in sustaining fisheries vitality must also be cardinal (think here, for example, of the importance of menhaden in fishery ecology). The earlier example of my angler self-assessment might be a method for initiating such thinking. Equally as important is the sustainability of the culture of recreational angling and the industry that supports it. The health of the culture and industry depends on the sustainability of the fishery, on the ecological awareness of the recreational angler. Culture and industry rely on the health of the ocean and fisheries, establishing a cyclical need for a healthy industry to support a healthy ocean and a healthy ocean to support a thriving industry. Moreover, we must now acknowledge that the sustainability of the ocean depends on industry and culture, as well. Thus, we need industry leadership in promoting not just a healthy ocean but also a new angler's ethic that protects fish and ocean. Changing that cultural ethic is foremost in protecting the future of the angling industry and culture, as well as of the ocean itself.

For the angler who only navigates, who misses the importance of the waypoints along the way, there is little connection to the spaces between point A and point B. In fact, we can look at the history of Western civilization as a never-ending philosophical and technological push to improve how we move from point A to point B. Our movement is designed to reduce the amount of time any such transposition requires and, as such, renders imperceptible the places and spaces that fall between A and B. Think, for example, about the characterization of "fly-over states": states that we only fly over when moving between airports in the east and west. The term suggests how we know these states, how we see these states. Travelers who fly over them only ever see them from above, from the surface, never learning about the territory itself, the inhabitants, the culture, the ecology. Navigational transportation strives to remove the very idea that there is space between point A and point B, rendering the places in between invisible, distracting our view from those spaces. The wayfarer makes more visible, more apparent the spaces across and through which he or she travels.

We live in the age of binge. Binge watching, binge eating, binge drinking, and binge fishing. It is not a new thing; binge has evolved over a good deal of time. Many of the novels we read as "classics" are the result of binge mentality. Writers such as Charles Dickens, for example, published many of their novels in magazines as serials. It was only after readers expressed frustration about having to wait for the next issue of a periodical to get the next installment of a story that editors started packaging all the episodes together in a single volume so readers could move from beginning to end without the wait. We have grown accustomed to the binge. Cramming activities into limited times and spaces. Ours is a culture of arrival and completion, product over process. We want to arrive at our destination; we want to be finished with a task; we want completion. We do not want to wait a week for the next episode of a television program or installment of a story. We want to be entertained now, and we want the satisfaction of knowing how things turn out. We desire the closure of the end. I wonder sometimes if we invest more in having fished than in preparing to fish. To say "I went fishing" swells with the expectation of memory and experience, victory and fulfillment. To say "I will go fishing" can only ever carry the weight of anticipation, never the promise of consummation.

This is a difficult atmosphere in which to be an angler, to participate in a pastime that has historically promoted and required slowing down. The allure of angling has always included the opportunity to exhale and be deliberate. Fishing, we are told, requires patience, but patience is not a virtue in an age of binge. This most contemplative of pastimes exists either as escape from the age of binge (that race of rats) or in direct conflict with it.

The FWC survey makes evident that our national culture of urgency has pushed our thinking about recreational fishing to genuflect toward binge. Rapid movement from point A to point B is anticipated in the same way that we anticipate rapid food delivery, instant communication, the next episode. The assurance of completion measured in the taking of fish is akin to the comfort of knowing the story will end in only a few more episodes that we can download right now. We move from point A to point B on the water, from dock to fishing site along charted, linear routes that mimic our terrestrial, road-bound navigation, venturing neither off road nor subsurface because our very ability to think about movement remains directive, pushed by the need for rapid transit and binge fulfillment. Our failure at such results in road-rage-caliber frustration (ever been to a boat ramp late in the evening on a holiday weekend?).

Do not get me wrong; this is not a condemnation of binge culture—particularly that promoted through the digital. I am deeply invested in it from my television- and movie-watching habits to my research habits, which thrive on immediate access to information and the ability to circulate information quickly. Likewise, the ability to rapidly download weather information, fishing reports, how-to videos, and other fishing-related information certainly enhances my life as an angler. There is, however, significant difficulty in binge, in an ever-growing need for increasing speed: with speed comes friction. As Ingold explains, "Unlike the wayfarer who moves with time, the transported traveler races against it; seeing in its passage not an organic potential for growth but the mechanical limitations of his equipment."[7]

The quandary of transport versus wayfaring and the culture of binge is palpable in tournament fishing, when time limits to fish are compounded beyond average restrictions such as work and life to include tournament-imposed restraints. For those who fish tournaments, the sense of urgency can become overwhelming, creating an atmosphere of pressure and desperation. The drama of boat-ramp urgency can astound on any given day as boater-anglers grow impatient waiting for others to offload or load boats, frustrated with ramp-based delays slowing their own progress. The ramp and other boaters form an imagined impediment to the angler's ability to arrive at point B. In such imagined urgency, the need is to not be waiting and killing time at the ramp but instead to already be en route to point B, where the fish are inevitably waiting. Add to this urgency the competitive edge of the tournament and the potential to win prizes and cash, and the need to be at point B becomes unbearable. The industry, of course, capitalizes on the craving for technologies to surmount transportation delays.

I itch for the adrenalin of the tournament, the caffeine-riddled anxiety to get everything just right. I revel in making fishing competitive, whether a simple wager of two friends fishing together or a big-money offshore tournament. However, even in that human-versus-human clash, it is the human-versus-fish dynamic that drives my passion, the angler passion. In those moments of hook and line and fish, the urgency, the binge, the competition slide beneath the surface, the angler left in harmony with the moment. These precise junctures in time and place disrupt the progress from A to B, compelling the angler to concede connection with the immediacy of the location not as point B but as an unscripted instant of habitation, of being in that place. The angler ethic emerges in such moments of place as bound not only to the fish at the end of the line, but also to the very place inhabited at that moment by fish and angler alike. It is here in this place, guided by such an ethic, that I long to be lost.

I imagine being lost at sea. Not located at point A or B, but somewhere uncertain between, or nowhere near A or B. We frequently attribute characteristics of disorientation, anxiety, worry, perhaps fear to the state of being lost. We generally do not want to be lost; we get lost accidentally, rarely on purpose. However, to be lost may also suggest being off track, off course, our direction misplaced. In such a state, we may be lost in thought, contemplative, and faraway. Lost at sea, my mind is absorbed and captivated by the oceanic. Often we hear the angler described as contemplative and meditative, descriptions that are synonymous with being lost in thought. Being lost at sea allows the angler's ruminative disposition to dissolve into the lost-ness, the condition of being lost. Such is not a loss, but a lagniappe, one perhaps best alluded to in Henry David Thoreau's overquoted truism: "Many men go fishing all of their lives without knowing that it is not fish they are after." Often we fish to lose ourselves, to be lost from the everyday. We breathe deeply the ocean air, trying to exhale the ordinary and willing ourselves to be lost in the moments of marvel we find in our time on the water.

Many years ago, perhaps as long ago as a quarter century, I ventured with a friend to Alaska for the first time. The trip had such a profound effect on my fishing life that I return to Alaskan waters as often as I can. That first northern exposure awoke me to halibut, salmon, grayling, lingcod, and cutthroat, fish I would never know in South Atlantic and eastern Gulf of Mexico waters. In the few weeks we were there, with a small group of my buddy's friends, we made a multiday kayak excursion along Resurrection Bay south of Seward. Stealthy ocean kayaks on big, mountain-lined water accented with spectacles of calving glaciers, my perception nearly overwhelmed. The depths of blues, both oceanic and glacial, transcending the visual, exposing the emotional. The panorama literally breathtaking.

Toward the middle days of our venture, I trailed behind the other kayaks, lackadaisically trolling a diamond jig behind me, hoping to pick up a small halibut or sea-run salmon for the cook fire we would build that evening on a mussel-lined beach. I was as lost at that moment as I have ever been on the water. I gazed at cobalt-gray water, lost in the reflection of the sun that peeked from behind swift-moving clouds. I watched enthralled as puffins, surf scoters, and other seabirds swam and flew about.

I reveled in the sight of harbor seals and sea otters that cautiously popped their heads from beneath the water to inspect my kayak and me. I took pleasure in seeing the bergy bits and growlers drifting by me in the slushy water, imagining I was paddling in an enormous blue curaçao margarita. Each moment an epiphany of discovery; the moments melding together and saturating my senses. My immersion beyond description.

My jubilation, then, interrupted when our guide, my buddy's best friend from college, paddled back to me, hollering that I needed to get a move on if we were to make our targeted landing point in a reasonable amount of time. Moving on, I mumbled, might be the most insulting thing we could do in that place at that time. I wanted to move with the place, not on it or from it. I wanted to be in that place and nothing more. Moreover, I certainly did not want to be disrupted from the solace of being lost. He had no response but to turn around and paddle on.

Our guide's response to my wayfaring reeked of impatience. His remarks, though, were neither personal nor particularly impassioned. They were merely conventional. Even if our travel plans were assumed and unspoken, implicit in the very idea of an excursion, to deviate signified intrinsic contrarianism. Objectives and destinations dictated our trip. My temporary loss of purpose slowed progress, and that is downright un-American. Onward. Personally, when it comes to fishing and time on the water, I canter toward thinking like Frank Zappa, who said, "Without deviation from the norm, progress is not possible." It is not a unique quote; plenty of people have argued that deviation drives progress and, certainly, encourages understanding. In addition, some, such as Robert A. Heinlein, have said so brilliantly: "Progress isn't made by early risers. It's made by lazy men trying to find easier ways to do something."

I deny neither my laziness nor my deviance. Both have

served me well in my fishing and my life. Such characteristics (or are they flaws?) have pushed me to take my time on the water, to observe, to stray from the plotted course, to venture farther out. In these ways, I find wayfaring to elevate my ability to be lost at sea and the opportunity to become more aware of my oceanic surroundings. Being lost helps me see. A successful angler must learn to see and to read in marine environments differently from how he or she sees elsewhere. The wayfarer learns to see more fluidly, in a way not learned in a world of print and two-dimensional image. Fluid sight, fluid reading, teaches us to see with the water, not to simply look at it. The wayfarer's vision incorporates the perceptual, the ability to perceive difference within the chaos of fluidity. Our sight moves from the vast expanses before us in broad view, to a rapid pinpoint focus when the slightest deviation appears in the ceaseless motion of water, wind, and sky. We learn to look for anomalies. We move our focus from the expanse to the details. We might say of wayfaring the same that hunting ethicist James A. Swan has so eloquently explained about seeing and hunting: "When you suddenly switch from seeing patterns to looking at details, seeking what Zen meditators call 'one-pointedness of mind,' you have the greatest chance for accuracy."[8]

The surveys stopped coming. I'm not sure whether their termination was the result of the answers I provided not being of value to the researchers, or whether they simply decided they had enough data to initiate analysis of what they had gathered from all those who complied and responded to their questions. Our point As and Bs entered into the archive, assessed and reviewed. I can only hope that whatever the researchers learned will benefit long-term oceanic and angling sustainability and that my own angling benefits from my time somewhere between A and B.

8

The Flying Car and
the Upgrade Path

Though often described, the future is always
unknown; its characteristics can be imagined but
never observed.

—James A. Herrick, *Visions of Technological Transcendence*

I first encountered the word "wondrychoum" in Mark Kurlansky's *The Last Fish Tale* (2008), and though not the word
itself, the concept was in his book *Cod: A Biography of the Fish
that Changed the World* (1998) as well as in his article "A World
without Fish" (2011) and his hybrid graphic text *World without
Fish* (2011). The word is unusual in its sound and its use. It
sounds outdated, a thing of the past. According to Kurlansky, a
wondrychoum was an early beam trawl device, a net suspended
from a beam, weighted and dragged along the bottom of the
ocean. The wondrychoum was designed to harvest greater
quantities of North Sea fish stocks more efficiently. According
to Kurlansky, the English had begun using the wondrychoum
and other beam trawl variants as early as the fourteenth century. The wondrychoum was limited in its efficiency because
the sailing vessels used to tow it along could pull only so much
weight, thus limiting the size of the nets that could be pulled.
For the device to work, the boat would have to pull the net
faster than the water current to ensure that the net dragged
behind the beam. Kurlansky shows that these early beam trawls

could sweep only about two miles of ocean bottom in an hour, limited by the technologies that powered the vessels, caught in a formula that pitted drag and weight against limits of speed and power.

Nonetheless, the wondrychoum changed global harvesting methods with such efficiency that we can now point to it as an identifiable techno-historical moment that profoundly influenced our current fisheries management problems. Early English fishermen, in fact, anticipated the effect of the wondrychoum and petitioned Parliament to ban such technologies. Kurlansky writes:

> But in 1376 fishermen petitioned Parliament to ban it, or at least increase the size of the mesh, because it swept up fish indiscriminately, taking many undersized young fish. Though it was not banned, fishermen in Cornwall and Scotland continued to denounce this type of gear, insisting that their hook-and-line technique avoided bruising of nets, thus harvesting higher quality fish. But they also insisted that the beam trawl damaged spawning grounds and would decimate fish stocks.[1]

Such petitions continued into the seventeenth century, but according to Kurlansky, "by 1774, beam trawling had become one of the principal fishing techniques in the North Sea."[2]

Kurlansky's rigor in presenting the history of harvest technology across his work is commendable, and his writing about cod and commercial harvest promotes a critical conservationist position. His accounts of the wondrychoum are similar to many other histories of technological innovation, tracing similar chronological paths. However, what Kurlansky does not address is how the wondrychoum is indicative of a cultural approach to technological innovation that sits at the heart of all ocean conservation and must be central to the new ethic of the recreational angler.

As I noted, Kurlansky explains that the wondrychoum was limited in its effectiveness because the associated technology of ship locomotion could not provide sufficient speed to pull larger versions of the wondrychoum. Trawling vessels just did not have enough power. However, in the simple acknowledgment that larger trawling devices would require more powerful vessels, we see our technological conundrum: we identify a problem or need, and we work toward a solution and eventually overcome the technological lack. There is a cliché that I find useful in thinking about technology: when we invented the car, we also invented the car crash. This stale saying makes evident that whenever we invent something, we also invent the problems associated with that thing. More telling, though, is that this saying gives direction to how we think about technology. We might revise this statement, for example, to say that when we invented the car, we also invented the flying car. That is, our technological advances traverse two inextricable paths: one marked with caution regarding the dangers of the invention and the other encrypted with the possibilities of what comes next.

Certainly, then, when paleohumans "invented" fishing, and even commercial fishing and aquaculture, they also invented overfishing and a host of other problems associated with ocean harvest. But they also invented fishing's flying cars.

A student once asked me, "Why don't we have flying cars yet?" To which I replied, "We do." You see, technology can be thought of as unfolding in two interlocked paths: one we might describe as theories of technology or the technological imaginative. The other might be the material manifestation of those imagined technologies. That is, if we can imagine a flying car, then it is likely we can imagine how we might actually make one. In many ways, then, the technological imaginative not only outpaces the material development of technological artifacts; it drives that second path,

Think, for example, about Italian electrical engineer Guglielmo Marconi, who is often credited as the inventor of the radio. When Marconi began theorizing the very idea of radio waves, his theories served no immediate material need. His technological theories needed to evolve and find the right moment and context to emerge in the form of a functioning radio transmission device. That is to say, when we imagine a flying car, we invoke it into the cultural technological imaginative, rendering it no longer simply imaginary, but part of a larger technological possibility.

If we are to speak of overfishing, climate change, and pollution as the three primary culprits of ocean and fisheries devastation, then we must speak of technology in the same breath. "Technology," of course, is a deeply problematic term, and in its most rudimentary understanding is reductively cast as either that which will be the demise of all that is natural or that which will be the savior. For the recreational angler to develop a new ethic that accounts for the future of the world's ocean, technologies of fishing and their contributions to marine devastation and reclamation must inevitably play a significant role. We must think not just about rods and reels, boats and ships, hooks and lines, fiberglass and wood, petroleum and ice, but about the very fact that fishing itself is a technology that emerges in various historical moments and cultural formations in a variety of ways. We must examine not only the technologies of fishing, but also the very ways in which technological developments have constructed and reconstructed our understanding of what ocean is and how we can interact with it.

In a 2008 interview in the *Guardian*, British scientist James Lovelock—who is best known for proposing the Gaia hypothesis, which posits that the Earth's biosphere is a single self-regulating organism—argues that global warming has already reached the tipping point and that catastrophe is inevitable. He goes on to say that "the sustainability brigade are insane to think we can save ourselves by going back to nature; our only chance of survival will come not from less technology, but more."[3] He argues for technological advancements such as nuclear energy and synthetic food. I find Lovelock's argument intriguing, particularly in terms of the future of the ocean. To be blunt, ocean conservation movements do not do enough in the face of global politics and attitudes toward fish and fishing. The future of the ocean lies not in renaturalizing the ocean, but in aggressive technological intervention: fish farming, genetic modification, nanotechnologies and biotechnologies for reducing pollutants, cloning and breeding programs, and so on. These may sound to some like science-fiction solutions, but they are technological possibilities, significant parts of the technological imaginative that we should pursue in order to build the ocean of the future in ways that can provide nourishment to the world's population while also prescribing a culture of protection of the ocean's diversity. If technological development continues to enhance both sportfishing and commercial fishing, making harvesting more efficient, then conservation technologies must outpace those technologies. Only within a framework of a culture of responsibility will we be able to negotiate the need for technological advancement as a mechanism for ocean conservation.

To these ends, the mind-set of the recreational angler must

change, too. We cannot simply defer responsibility to the idea that technology will save us from our predicament. New baselines for recreational fishing cultures must promote new ethical practices toward the end of ocean conservation and fisheries sustainability in ways not addressed before. The size, numbers, and regulations of sportfishing must reflect new conditions.

The political battles over harvest "rights" must also be redressed culturally. The ocean cannot be regarded as a free-for-all zone where federal, local, and international regulations are seen as impingements on individual rights. The public trust is not open to the individual; it is open to the public. This includes the argument of the commercial industry that it provides access to ocean resources to the public for whom these resources are not easily accessible geographically. These are not "resources." The global protein problem can be addressed only by farming, synthesizing, cloning, and genetic modification technologies if we assume that fish protein will continue to play this role in the global protein economy. The idea that all of nature might remain natural is no longer a viable lone mind-set. If we are to preserve any of that nature, then synthetic nature must provide for the masses what has previously been supplied by the natural world. We can no longer have one without the other.

Technology is cast as a cause (if not *the* cause) of environmental destruction, but such scapegoating is too easy, too simplified. The necessity of rapid, radical technological development now faces us. The environmental-technological conflict has already been superseded by the environmental-cultural conflict. The new recreational anglers' ethic, then, needs to imagine a future ocean of hybridity between technology and nature. Ultimately, the new anglers' ethic must evolve from an amalgam of influences: recreational fishing culture and tradition, recreational fishing industry influence, science, technology, and the unnamed spirituality of the human-ocean interface and interconnectedness. In many ways, this amalgam already drives recreational angling culture, though often in subdued ways. We now need to recognize that fishing itself is a technological intervention, that without some form of innovation—be it Paleolithic fish traps or thermoplastic elastomer lures—the very act of fishing would not be possible.

Likewise, we must acknowledge that our development as anglers, as an industry of anglers, as a culture of anglers, is embroiled in our technological evolution. We can point to technogenesis—the coevolutionary process of development that coils technology and human together in symbiosis—as integral to the emergence and sustainability of recreational angling. I suppose an easier way to say this is that recreational anglers love gear and tackle and such, and without it, out fishing culture would not have advanced economically or culturally the way it has. We have to admit that to be a recreational angler requires that we be beholden to our technological developments to some degree. This degree, however, may be personal, or it may be institutional. Making such choices—or, more appropriately, making such choices evident—will be part of the new recreational anglers' ethic.

A few years back, one of my fishing mentors called to see whether I wanted to make a short run from his dock across the inland waterway to pick up a few redfish for dinner. I wasn't about to turn down the offer, so I grabbed my gear and drove over to his place, where we jumped into his Hewes and made the short run across the water. When I climbed onto the boat that day, I was holding a tackle bag packed with two or three satchels of lures, a satchel of hooks and other terminal tackle, three or four rods rigged and at the ready, wading boots, bottled water, handheld GPS, rain jacket, bait bucket, aerator, and an assortment of other stuff. My mentor had one rod, tied with one simple red-and-white DOA Shrimp. Nothing else. He looked me up and down and asked, "You got everything you need?" suggesting I might be carrying too much gear, though I didn't catch the dig at first.

"Yes," I replied, "how about you?" He held up his one rod. "That all you have?" I asked.

"What else would I need?" he responded stoically. Moreover, he meant it. His post-Depression childhood had imbued a frugal mind-set, one that empowered him against waste and excess and the culture of binge. Spending $29.99 to download an entire season of a television series was incomprehensible, as was spending $15.99 for a lure or even having cases of lures when all you needed was one. From my perspective, his tackle was meager and in need of upgrade. My pretentiousness was driven by the angling industry's culture of urgency and mantra of upgrade. My tackle box wasn't full of just tackle. It was full of the newest tackle; it was full of a cultural promise that such tackle would provide the most direct access to catching the fish about which I dream.

One would think I would learn my lesson about the hubris of gear-tech and the modesty of experience and knowledge. It could not have been long before that day on the water with my mentor that I had gone out to the trap and skeet range to shoot a few rounds of sporting clays. There was an older gentleman on the range shooting alongside me. He wore labor-faded jeans, a flannel shirt, and boots worn with age and work. His gun was piecemealed together; the stock looked liked a mishmash of 2 x 4s and duct tape. The barrel corroded. The action pump, unusual for shotgun sports. When the club official called together a squad to the first station, I joined the group, as did the old man. The rest of the squad meandered out from the clubhouse, toting their Krieghoffs, Guerinis, Perazzis, and Benellis, guns whose names sound more like racing cars (and carry equivalent price tags). They wore shooting shirts and shooting jackets emblazoned with stitched logos from Beretta, Orvis, and the like. Their boots were closet furniture, not daily drivers, scuffed only from their climbing in and out of new-model SUVs. They snickered among themselves as they nodded at the old man's gun, congratulating each other on their gear-assured victories.

Needless to say, the old man was a crack shot, scoring a perfect a fifty-target round, besting every one of the rest of us easily. I will never forget that old man's knowing smirk at the end of the round as he sidled back to his beat-up pickup truck. No doubt, he showed us whippersnappers a thing or two about shooting and assuming.

I suppose I did forget the lesson of that day, though, because sure enough, my fishing mentor outfished me all afternoon. He put that DOA through a workout, pounding fish to a count of about four to one of my fish.

I learned two other things that day, lessons I still need to learn over and over again. The first is a lesson about access. Without the technology to travel quickly over relatively large spaces, most of my fishing would never happen. I depend on the automotive industry to get me to the dock and on the boating industry to get me to the fishing site. Ultimately, my fishing depends on petroleum technologies, and a host of other sciences and technologies that make it possible for me get to the water or on the water. I often wonder how the recreational fishing industry might have unfolded if such forms of transportation had not evolved. Would there be a recreational saltwater fishing industry if only those within walking distance could access saltwater, and how might that community of anglers engage one another if boating technology did not provide access to water beyond the shoreline? Similarly, how much less would we know about the ocean and fishing without those technologies? Access is, of course, privilege. My access is bound to my ability to afford transportation, to afford the time to slip away from daily necessities. The privilege of recreation is provided by a host of technologies that make access available, and it is also a privilege to be able to afford those technologies and the time in which to use them. Just as we talk about the digital divide and the effect access has on education and commerce, so too must we acknowledge the effect the technological divide has on angling access.

Second, I realize that though it takes some technology, some gear, we do not always need as much stuff as we think we do. This one is a hard lesson for me to learn. I love the gear, the tackle. I have shelves of it, storage bins filled with lures, lines, hooks, reels, and other accouterments of the pastime. I own gear for pursuing fish I will never pursue, simply because the tackle is beautiful or intriguing. I collect tackle as some might collect art (so much tackle is, in fact, art to me—or at least artistic). My wife often asks of my gear purchases, "Is that something you need?" I am always dumbfounded by the question, or by the very idea of the question, a question for which there is always already only one possible answer no matter the object about which it is posed: yes. I suppose I am the child to which Orwell's George Bowling refers in *Coming Up for Air* when he wonders, "Is it any use talking about it, I wonder—the sort of fairy light that fish and fishing tackle have in a kid's eyes? Some kids feel the same about guns and shooting, some feel it about motor-bikes or aeroplanes or horses. It's not a thing that you can explain or rationalize, it's merely magic." That's me: a kid in a tackle store.

Every time I fish, I worry that the thing I do not bring with me will be the very thing I need most, so I carry too much.

I am incapable of showing up to fish with one rod and one lure. I need the security of selection. I need the absolution that tackle grants me; it affords me the rationalization that it is not my skill (or lack thereof) that caused me to not catch fish, but the fact that I had the wrong lure or wrong line or wrong location. Just like the neophyte gamer's strategy of blaming the equipment—the joystick is not working!—tackle lets me blame something other than myself. Tackle is our never-ending quest for a panacea—or scapegoat. I see tremendous value in the role the recreational sportfishing industry plays in maintaining the culture and in forwarding the technological evolution, all in partnership with an active conservationist agenda. I find myself, then, in a conundrum of excess, curious about how much is too much, curious about whether I buy gadgets and gear for their actual contributions to my fishing practices or fishing ability, or purchase them as tokens and totems of my membership in the recreational fishing tribe or as excuses for my own failures.

I acknowledge that my participation as a consumer in the industry simultaneously makes me complicit with all that is problematic with the industrialization of recreational fishing and all that is generative. The good, the bad, and the ugly. I suppose that part of the evolution of the new recreational anglers' ethic requires the recognition—if not the confession—of our contradictions, individually and culturally. The willingness to identify that recreational angling is not always contemplative connection with nature, but that it is embroiled with destructive elements and acts, as well. Recreational anglers embrace the technological benefit the industry provides us toward the goal of successful fishing, but we must also acknowledge and accept the responsibility for the ecological impact of such technologies and adjust our practices accordingly. I again note the example of the deep cultural shifts that recreational anglers made in order to adjust to the catch-and-release mind-set and practice. Similarly, recognition of the effects of discarded monofilament fishing lines led to cultural expectations of line disposal practices. We now face a host of other necessary cultural adjustments in light of increased awareness of the effect of angling technologies on the world's ocean. Foremost among these is the recognition of the global impact of plastics on the ocean.

There is no question that recreational anglers invest tremendous faith—and capital—in plastics. A quick glance at the fishing lure industry alone reveals plastic devotion. Plastic, which is an umbrella description of many forms of synthetic petrochemicals, permeates nearly every facet of our recreational fishing lives. We are culturally dependent on plastics. Certainly, we could fish without them, but we would surrender so much benefit in doing so. We could not, as well, surrender the mythology of the success we gain by their presence. Thus, our dependence is ideological and cultural, not latent.

Part of my fascination with the technology and gear of this pastime is my desire to understand the histories of many of the technologies we rely on, or that we tell ourselves we rely on. For example, the development of artificial lures intrigues me, particularly those that work to imitate dynamic details of the imitated bait species. Shrimp are one of the most frequently mimicked forages for inshore anglers. The saltwater lure industry has designed and manufactured countless variations of artificial shrimp since the 1920s, and understanding the history of that development provides insight into how our use of technology has driven angling efficiency and, perhaps, reliance. Given the complexity of the parts and locomotion of shrimp, developing artificial shrimp requires significant attention to detail. Anglers, artisans, and engineers have been trying to manufacture suitable shrimp mimics for more than one hundred years. In 1928, George R. Ettles, an angler in Jacksonville, Florida, applied for a patent that was granted in February 1931 for what he described as "a novel lure . . . composed of transparent material and an assemblage of parts made sufficiently in imitation of the natural shrimp to make the lure effective in attracting fish." In 1929, Ettles's company, the Florida Artificial Bait Company, began making the Superstrike Shrimp based on this design. The story of the design of this shrimp is part of saltwater fishing lore, as Ettles designed and made the first Superstrike Shrimp from the wreckage of an early human-powered flying machine that his brother-in-law Henry Farrin bought from Captain George White of New York after Captain White crashed the machine while attempting to fly it on St. Augustine Beach. The Superstrike Shrimp was fashioned from the celluloid material of the flying machine's wings. As you can see in the schematics included here from Ettles's patent application, the Superstrike was a complex lure with many moving parts dependent on technologies from other industries, primarily aviation.[4]

In 1949, Nick and Cosma Creme, residents of Akron, Ohio, developed the first mass-produced soft-plastic lure, the Creme Worm. Using a mix of vinyl, oils, and pigment, the Cremes began manufacturing the worms and selling them by mail order in 1951 for one dollar per pack of five. Interestingly, the Cremes' business grew alongside the southern sport of bass fishing in the early 1950s, and the demand for the Creme Worm rapidly increased. The worm was such an effective bass lure that the state of Texas temporarily banned its use. More importantly, the Cremes' process for molding soft-plastic lures opened the door for other manufacturers to experiment with soft-bodied forms other than the worm, including shrimp.

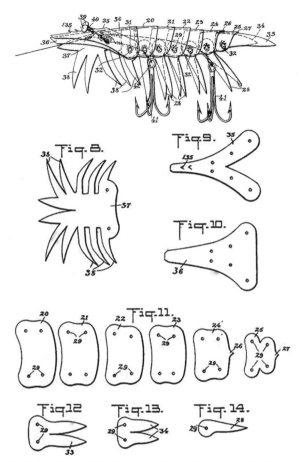

Details from George R. Ettles's patent application for the Superstrike Shrimp

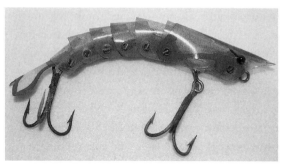

George R. Ettles's Superstrike Shrimp

Given the ubiquity of the shrimp as an inshore forage and its popularity as bait for anglers, many companies began experimenting with developing a shrimp that could be molded with soft plastics and still be durable enough to withstand the abuse of bites by inshore fish species. Ultimately, Mark Nichols's DOA Shrimp would rise to become the soft-body standard in the realm of artificial shrimp. DOA founder Mark Nichols patented his lure design more than twenty-five years ago, and just about anyone in the industry will tell you that the DOA Shrimp design is the gold standard. I suppose in the lure industry, imitation is the sincerest form of flattery, because Nichols's design has been imitated many times. Nichols started designing shrimp lures because he was fascinated with the way shrimp move; he spent a good deal of time observing shrimp and wanted to make a lure that would drop the same way a shrimp does. His first designs did not quite achieve the objective, but he refined them to focus on the movement of the swimmerets, designing a unique internal weight system and experimenting with different colors. Nichols's more recent patents reflect his innovations and dedication to perfecting the artificial shrimp. I am intrigued with the hybrid science Nichols used to construct his lure patterns. His work combines an understanding of plastics and molding systems with careful long-term analysis of shrimp biology, behavior, and ecology.

By no means, though, did Nichols's study of shrimp and production of artificial shrimp solidify the approach to manufacturing artificial shrimp lures. The development of thermoplastic elastomers along with the emergence of 3-D scanning and printing technologies allowed companies like Savage Gear, Live Target, Egret Baits, Monster X, and others to begin to produce remarkably rugged, durable, and lifelike plastic lures. Likewise, since the 1980s, many lure manufacturers have developed processes for infusing scent additives into the plastics during the molding process, creating natural-looking and natural-smelling lures—Berkley Gulp! stands as exemplary in this category. While we may look to an artificial shrimp as fulfilling a rather simple industrial and cultural role in salt-water angling, the shrimp's technogenesis continues to affect angling practices. Likewise, while we can point to continual technological development as benefiting the fishing industry and ultimately the angler's repository and perhaps skill range, we are bound within the new recreational anglers' ethic to question the effect of adding new plastics to marine ecologies.

Like many who fish in saltwater, I feel compelled when on the water to pick up the floating trash I find. Sometimes I overfill my kayak with trash, making actual fishing difficult. Other times my boat looks like a trash barge. I am always amazed by the volume of trash I pull from the water and by the array of debris I find. Unquestionably, among the usual suspects of beer bottles, plastic water bottles, chip bags, and so on I also regularly find discarded fishing tackle. I would bet that just about anyone who fishes in saltwater regularly, or walks along a beach or dock, finds discarded—intentionally or accidentally—fishing tackle, lures, and hooks in particular. Years ago, a friend of mine began collecting every discarded lure he found while out fishing. At first, he just kept the lures in a pile, but as his collection grew, he began hanging them on a board in his house. Some of the lures were virtually new; others were corroded with time and saltwater. He hung them all, and as the board began to accumulate hundreds of lures, those of us who visited his house began to see the collection as a piece of art. It was beautiful not only as an aesthetic piece composed of a kaleidoscope of colors, but also as a polychromatic commentary about waste. The piece, which continues to grow, stands as a testament to recreational angling's complicity in a culture that contributes to the discarded plastics in the ocean. Of course, other anglers who have seen my friend's "art" ask whether they can have some of the still-usable lures. Some ask why he does not just throw away the lures that are no longer of angling use. I have heard him explain that for him, each lure carries the same value—the ones some see as still usable and the ones they identify as having no angling value. To my friend, they are all reminders of a more holistic issue regarding anglers' excess and anglers' own faults in ocean desecration. The collection, that is, reminds my friend of his own anglers' ethic.

I fantasize from time to time about various ways to challenge my fishing. For example, I consider fishing with only bucktails for a year, or only spoons, no matter the target species or location. Could I have a successful year relying on only one of these two canonical lures? Or, how long could I fish with just one lure? Tie it onto my line on January 1 and not switch it out until it breaks off; how long would it last? How many fish would it catch? Or, what would it be like to fish with only the tackle in the tackle box I inherited from my grandfather when he passed away? Or with my first tackle box, which sits in storage? How long I could fish with what I already own? (Though this is an unfair challenge as I have a fairly substantial cache, and I would have to surrender reviewing new gear for Inventive Fishing, which I'm not willing to do.)

Of these challenges, though, the one I find most intriguing is fishing only with scavenged terminal tackle I find on piers, along beaches, snagged in rocks and coral, or tossed aside in the water. That is, fish only with what I find expelled into the ocean. Instead of mounting the found lures as trophies, as my friend does, I contemplate recycling them in order to maintain their intended use-value, thereby not condemning them to the status of debris or even art. Let them catch fish again. Given the array of discarded terminal tackle I regularly find tossed into the ocean, I presume I would manage just fine. In fact, I would put it to all recreational anglers—saltwater and

fresh—to start your own collection of found tackle. Not only
would such collections contribute to conservationist anglers'
efforts to remove trash from our waters, but I'm betting they
would teach us each a good deal about how we function as
anglers: what gets discarded in a specific region, which lures
are common to an area, which lures degrade more readily, and
so on. It might be a compelling cultural shift if, alongside stories
of our greatest fishing moments and pictures of our trophy fish,
we were proud to show off our collections of reclaimed tackle.
Ponder, that is, a cultural practice of lure reclamation apace with
catch and release, line recycling, and banana aversion. Cultural
nuances like this contribute to the overall new anglers' ethic
and our willingness to see the details within the complexity of
recreational fishing.

The new anglers' ethic, though, requires asking questions
that are more complex than what we put in the ocean and what
we take from it. Many of these questions will manifest only as
our technologies develop. Others will require intellectual and
cultural shifts before we can identify our questions yet to come.
Inevitably, though, given the history of our relationship with
the global ocean, many of these questions will be questions of
adaptability—the ocean's and ours. Questions that ask about the
ways in which the ocean and its inhabitants will adapt to both
our additive and removal practices must become paramount.
For example, in 2017 Representative Jaime Herrera Beutler
(R-WA) introduced H.R. 2083, the Endangered Salmon and
Fisheries Predation Prevention Act, which seeks to "amend the
Marine Mammal Protection Act of 1972 to reduce predation
on endangered Columbia River salmon and other nonlisted
species, and for other purposes."[5] In a nutshell, this act seeks to
revise the Marine Mammal Protection Act (MMPA) to allow
for the "harassment" and killing of sea lions (which the MMPA
protects against) in order to discourage sea lions from feeding

on endangered salmon populations. According to the proposed bill, California sea lion habitat was historically restricted to saltwater environments. However, as many as three thousand sea lions have recently been foraging as far as 145 miles up the Columbia River in Oregon, altering local ecologies. Similar patterns have been observed elsewhere in Oregon and Washington. H.R. 2083 identifies that "Federal, State and Tribal estimates indicate that sea lions are consuming at least 20 percent of the Columbia River spring chinook run and 15 percent of Willamette River steelhead run, two salmonid species listed under the Endangered Species Act of 1973."[6] Sea lions facilitate this recent behavior by exploiting human technological intervention in the rivers. Sea lions have begun using the dams along rivers to expedite their salmon hunting, including using "fish ladders," which were installed to encourage salmon migration and spawning. That is, the introduction of dams and fish ladders to the rivers has provided sea lions with the technology needed to access the salmon on which they forage. This adaptation has significantly altered the salmon ecology of the Northwest, changing predator-prey dynamics in ways that are having significant impact on fish populations, pitting the conservation value of one endangered species value against that of another. In response, Beutler designed her legislation to protect the salmon populations without considering such dynamics of the problem as the ecological impact of changes to sea lion habitat, the introduction of dams and fish ladders, and the possibility of actions against the sea lions without further study. In June 2018, the US House passed H.R. 2083. As an angler, I find myself philosophically supporting the bill, but to do so falls into the ongoing trap of making decisions based on ideology rather than on scientific evidence.

Unfortunately, we fall too often into this pattern. We also

must recognize that our technological evolution has contributed to ocean degradation for much longer than we assume.
That is, human development has impacted ocean conditions
for much longer than the immediate technological age or
industrial age. Callum Roberts has clearly demonstrated that
the assumption that human-generated ocean degradation is a
recent historical event is profoundly incorrect, and that human-
wreaked demolition began much earlier than we tend to
imagine. Similarly, we can show that the relationship between
human technology and the ocean predates even the wondrychoum and has had even larger global impact than the destruction of trawling and industrial harvest.

Consider this, as well: from 1854 to 1858, the Atlantic
Telegraph Company laid the first transatlantic telegraph cable
from Foilhommerum Bay off Valentia Island in western Ireland
to Heart's Content on Trinity Bay, Newfoundland. On August
16, 1858, Queen Victoria sent US president James Buchanan
the first transatlantic telegraph. Prior to the cable, transatlantic
communication relied on ships to carry information, but this
technological achievement altered more than just global communications. The initial cable routes, as Nicole Starosielski has
shown in her eye-opening book *The Undersea Network*, laid the
foundation for communications networks that have influenced
global communication, culture, geography, environment,
history, and politics. In fact, undersea cable networks are now
responsible for carrying almost all transoceanic internet content
and serve as the foundation for the functioning of wireless
networks (remarkably, despite common assumptions, they do
so more quickly, more cheaply, and more often than satellites).

In 1860, a damaged portion of the transatlantic cable was
lifted from the ocean floor nearly six thousand feet down.
Those working on the cable found numerous deepwater species

entangled along the line. Prior to that moment, conventional wisdom, supported by scientific observation (limited by the technologies of the time), understood the ocean to be lifeless below depths of about 1,800 feet. The discovery of deepwater life radically changed not only what we knew about the ocean but how we thought of it and positioned it culturally.

From 1928 to 1929, working with engineer Otis Barton, explorer and naturalist William Beebe built a bathysphere, an unpowered, pressurized submersible. Beebe used the bathysphere to make multiple dives to three thousand feet between 1930 and 1934, from Nonsuch Island in Bermuda. Beebe's bathysphere dives marked the first time humans had attempted to observe deepwater organisms in their natural environment.

As a kid, I was fascinated (and still am) with Beebe's books *Beneath Tropic Seas* and *Half Mile Down*, as I was with Jules Verne's book *Twenty Thousand Leagues under the Sea* (which, by the way, was originally serialized in the French periodical *Magasin d'éducation et de récréation* over sixteen months between 1869 and 1870 and then published as a complete work in 1871 when readers demanded the complete story—think binge). Verne gave us speculation, imagined visions of the ocean's depths. His imaginary ocean was spectacular and adventurous, but his technological descriptions were ahead of his time. He gave us the *Nautilis*, a remarkable submarine at a time when submarine travel was in its infancy, moving from the technological imaginative to the material manifestation. Verne gave us a flying car. Beebe, on the other hand, gave us a firsthand glimpse of the deep ocean, a place we could only imagine by way of writers such as Verne prior to the technological development of the bathysphere. In the same period as this technological development provided the first access to deepwater study, giving us first glimpses of previously inaccessible ocean regions, Ernest Hemingway was living in Key West developing the principles that would come to define recreational sportfishing ethics and practices.

What Beebe, Verne, and others also began revealing for us was that the sophisticated technology needed to experience most of the ocean's space would only ever be accessible to us through written description. The necessary technology would remain out of reach for most of us. In this way, too, the technologies required for deepwater exploration manifest in ways that most of us can construe only as a kind of technological sublimity. We gaze at a bathysphere or contemporary submarine and stand stupefied by the overwhelming sense of this technology's vastness and what it opens up before us. The technological sublime embraces the oceanic sublime. We drift above the depths, able to only imagine what lies beneath. Our imagination is inspired by words and images from those with technological access to give us glimpses, not always by our firsthand experiences. This is why Roy and I had to know that day on the Gulf what it felt like to swim in water that deep. That was a moment of cultural immersion, a moment to embrace the oceanic and technological sublime. Culturally we imbue these sublime moments with our inability to grasp the measurelessness of our interactions with the ocean and its inhabitants. The sublime—whether oceanic, technological, or other—is awash in astonishment and adorned with a tint of horror. We are overwhelmed by our oceanic technologies, so that many of the world's population can only ever try to grasp the sublimities by way of deeply mediated representations. Even television shows that project filmic representations of ocean depth can only ever contribute to the global view of the oceanic sublime and the technological sublime. Few will ever have the opportunity for firsthand experience; even fewer will ever be able to overcome the sublimity of the firsthand experience. For too many, the engagement with the sublime results in deep-seated, culturally circulated fear—what E. M. Forster would call the "terrors of direct experience."

The ocean needs an upgrade. As recreational anglers, we need to contribute to a careful, conscientious technological intervention at all levels that strives to weave a path for technological upgrade with a long-term conservation agenda. This will include not only careful consideration of the materials we use to fish, but global consideration of the technologies we employ to enhance fisheries for the benefit of the global protein economy. It will include focus on harvest concerns at a large scale, as in the massive concerns of deep-sea oil drilling and industrial fish harvest. It will concern minutiae such as the understanding and sustainability of deepwater marine microbes—perhaps some of the most important organisms in the scientific study of everything from climate change to biotechnology. The upgrade will also require a reconsideration of space, of our very understanding of the smallness of what we have assumed to be a vast and endless ocean, and the role of nanotechnologies in protecting that space.

Ultimately, as recreational anglers, our relationship with the ocean is inherently mediated technologically. No matter our beliefs that recreational fishing helps us connect with nature, and no matter the "get back to nature" sensibility we attach to our cultural promotions, no such connection unfolds without technological intervention. Unlike the timeworn causality dilemma "which came first, the chicken or the egg?" the question "which came first, the technology or the activity?" can be answered confidently by recreational anglers knowing that technology provided the access to the fish. Since long before Dame Juliana Berners wrote *Treatyse of Fysshynge wyth an Angle* in the 1400s, teaching anglers how to make essential fishing tackle such as gut lines, hooks, and rods, or before Izaak Walton, influenced by Berners's work, gave us the landmark instructional resource for fishing technologies of the seventeenth century, human technologies were already at work in forging our relationship with the ocean. Early harvest technologies that engineered our relationship with the ocean preceded the wondrychoum. As recreational anglers in a historic moment that requires the conscientious revision of our ethical engagement with the ocean, we find ourselves needing to turn our gaze to the technologies of our ocean pact. We need to question those technologies we have come to depend on—not to disavow their use, but to understand the ramifications and consequences of that use. We also need to acknowledge and embrace the idea—and in turn, act on the idea—that ocean sustainability will require not less technological intervention, but more. The new recreational anglers' ethic will need to embrace and promote technology not as the vilified cause of ocean destruction, but as fundamental to the future of the world's ocean.

9

The Knot

No angler merely watches nature in a passive way.
He enters into its very existence.

John Bailey, *Reflections from the Water's Edge*

I had been walking the beach since early in the morning, look-ing to the surf for signs of Spanish mackerel and bluefish as an easterly breeze picked up throughout the day. By midafternoon, the gusts had swung to the northeast and picked up substan-tially. The slate-colored water churned and chopped, frothing with ashen foam and climbing to iron peaks that smashed headlong into one another. This was angry water. It chewed at the beach, pulling out what it could and spitting ashore what it did not want. By reputation, this was the Graveyard of the Atlantic, the waters of North Carolina's Outer Banks, some of the most ferocious around. On this day, as it has often done since the barrier islands formed, the Atlantic turned dark and the wind bashed ashore, bringing squalls of rain and light-ning. I was not yet a teenager, and as the clouds consumed the daylight and the first drops of rain berated the shore, I could see my parents back on our cottage deck urgently waving me to come in out of the looming storm, a command to which I could not turn my back. For the rest of that afternoon, the storm pounded the beach, and I could only stare at its assault on the ocean from behind the cottage window.

In the late afternoon, with a consistent rain still falling and the sun making one last effort to penetrate the clouds, I kept watch over the surf through the window. It took a moment for my eyes to convince my brain, but there was something in the surf, something that the ocean appeared to be trying to rid itself of. A snarl of yellow, it seemed, that reached up and down the beach, wrapped in the waves, tangled in knots of seaweed, driftwood, debris, and foam.

I called my father to the window, wanting both confirmation that I was indeed seeing something in the low-light churn and some opinion as to what it might be. He was as confounded as I was. Rather than stare in confusion from the window, however, he had a fatherly solution to our curiosity: he told me to go out into the storm and drag whatever it was to the beach. He would keep an eye on me from the window.

So I did.

The ocean was ridding itself of a knot of rope—a massive ganglion of yellow, one-and-a-half-inch, twisted polypropylene rope. There was no telling how long it had been in the ocean. Not long enough to disintegrate into thermoplastic polymer microfibers, but long enough to entangle fishing lines and hooks, seaweed, and other debris, forming a floating yellow morass in a gray ocean. The ocean coughed and hacked, trying to expel this choking tangle.

Wading into the smashing surf, ducking under the crashing waves, I was able to get hold of one of the outreaching tentacles of the mass as it searched for more detritus to ensnare. At the risk of my own entanglement, I grabbed hold of the yellow sea monster and began dragging it to shore.

The tangled mayhem that I pulled to the beach that afternoon will be to me always the visual manifestation of what the word "mess" truly means. This was something no one should want to make. Nevertheless, in those first moments on the beach, this mess captivated me as something to unmake, a knot so problematic that I had to know whether I could untangle it. I needed to know whether it was composed of one rope or many. I needed to know what it had ensnared and enfolded into its mass, its contents clearly synthetic and biological, a writhing biotechnical gnarl. Untangling the yellow knot became a challenge I could not pass up. It was a puzzle, a problem to be solved, a trial to undertake. That afternoon, after I wrestled the gorgon to the beach, I worked my way back and forth along the ligature of the contortion for hours and into the evening. The rain kept the seaweed moist and alive in my hands, the rope thick and willowy. During the last glints of the day, I rinsed and coiled more than two hundred yards of yellow polypro line. For more than thirty years, I have made use of that line for purposes ranging from dragging felled trees to lashing ballast on boats. I have cut pieces from it of various lengths for varied uses. It is one of the longest-lasting material gifts the ocean has given me. As useful as that rope has been, though, it was the lesson of the knot that has been most beneficial.

Legendary (at least to those of us who fish with him and get to call him friend) Key West guide Captain Judd Wise (a.k.a. "The Shirtless Captain") attributes a good deal of his success as an angler to his ability to tie reliable knots. Captain Wise lives by and professes a simple rule about knots: tie them the night before you go fishing. His logic is sound: if you are tying knots while on the water, chances are you are not being as attentive to the knot as you would be when all you are doing is tying knots. As he puts it, if the fish are biting, you are thinking about the fish, not the knot, and you are likely to be in a hurry to get lines back in the water, rushing the knot. Captain Wise's logic is tantamount to the Scout motto: be prepared. Tying knots can be an art form, and the reliability of knots can be the difference between success and failure as an angler.

Of course, conscientious, experienced anglers like Captain Wise can usually tie reliable knots in just about any situation, but Captain Wise's night-before philosophy rings true for many anglers who are as attentive to rigging and knot tying as they are to any other aspect of their angling. Knots are a particular point of pride among serious anglers. A knot that fails while fighting a fish can be a heartbreaking disappointment akin to a kid who decides to attend Florida State. It can also be a public embarrassment if one's angling companions discover the weakness. Most of us would rather be mocked by our fishing buddies for having pissed our pants in public than for having tied an impotent knot—or for admitting our kid wants to go to FSU.

Knots can be paradoxical. They are at one moment artful and the next impossible. They range in use and task; they bind, connect, and secure. Knot theory conceptualizes knots that cannot be undone. Trefoil knots, the simplest of the nontrivial knots (knots that cannot be untied), for example, provide a conceptual loop, a never-ending knot without ends. The trefoil has appeared in iconography throughout history. It is also a knot in whose mathematical foundation we can read influences on the design of boat propellers. W. H. Stuart Garnett's 1907 book *Turbines*, for example, addresses the beauty of trefoil propellers.[1] It is at once a knot of perplexity and simplicity. For those of us with the propensity to tie knots—both metaphorical and material—and the proclivity to untie them, the trefoil confounds.

As a conceptual, visual analogy for the state of recreational fishing, we see the never-ending loop of culture, ecology, and economy knotted in never-ending tension and accord.

Trefoil knot (courtesy of Jim Belk)

A rigger's nightmare. The situation of recreational angling is now bound in a knot of culture, industry, ecology, and politics that perplexes. The anglers' new ethic is driven by a question of synchronism: how to both strengthen this knot in valuable ways while simultaneously untangling its cumbersome excess to the end of simplicity and strength—particularly when that knot may be a knot without ends, a knot, like the trefoil, which cannot be undone.

Recreational fishing has grown into a tangle, a knot of behemoth proportions: the clarity and ease of water, fish, human, and tackle have been twisted into a boundless (and expanding) contortion of regulations, politics, corporate economy, and global greed. The question for recreational anglers continues to be a question not of complacency and learning to navigate the snarl, but of learning to tug at the strands of its composition and retie it in a sensible fashion. Ultimately, this requires angler activism. This will require saltwater anglers to begin to recognize the strength of our numbers and the strength of the data that show the economic impact of recreational fishing on local and national economies. Our activism will need to be scientific in origin, prioritizing fact over belief, data over ideology.

Simplification. Untangle the knot. I have talked with many anglers who feel strangled by the recreational fishing knot. They tell me that all of the regulations, all of the politics, and all of the industry ruin the very idea of recreational angling. They claim that regulation and industry have nothing to do with them and their fishing. They are not interested in reading policies or letting legislators know their thoughts about regulations. They are angry with authorities for telling them when they can or cannot fish or what they can or cannot harvest. Some resist industrial or governmental intrusion into their fishing; others simply ignore it, remaining intentionally oblivious to policy and product. Many, though, echo a common theme among scads of recreational anglers: a desire for simplification. I know several anglers who profess a philosophy of simplification. They have rid themselves of their boats, opting for either wading or kayaking, wanting neither the cost of fuel and boat maintenance nor the responsibility for the environmental impact of petroleum use. Many have also divested themselves of excess gear, retaining only one or two rods and reels that have served them well. Some have even shed all the hardware, keeping only a few of their favorite lures. They do not want to hear about new lures, about innovations, about regulations. The industry-planned obsolescence of some gear frustrates them;

they do not want to buy a new reel each year. They do not believe that the newest lure innovation will somehow improve their chances of catching fish beyond what their current lure selections provide. Their upgrade path has leveled off. They just want to fish. Some even ignore regulations. Because they do not keep fish, they do not consider themselves as ever violating harvest laws (including closed seasons). Even those who do keep fish contend that the government has no right to intrude on their harvest rights. Some defy policies that limit their catch, labeling them as inaccurate, irrelevant, or an infringement. They fish where they are not likely to encounter wildlife officers. Some of the extreme antiregulation anglers refuse to buy licenses, citing licenses as a form of governmental intrusion on individual rights. Others identify their minimalism as a return to angling spirituality, a desire to be better connected to the places they fish, to the water, not to the materialist trappings of the industry. The materiality of angling, they charge, changes not just the fishing, but also the very essence of the angler.

I remember reading a political cartoon many years ago—perhaps it was in an outdoor magazine, but it was just as likely in a random magazine I picked up in a waiting room somewhere. The cartoon depicted two aboriginal men speaking to one another in the foreground. They were clothed only in loincloths, their feet bare as they stood on what appeared to be an island beach. To their side and a bit in the background stood a third aboriginal man decked out in waders, a fly-fishing vest, a creel, a hat lined with flies, a landing net, and other trappings of the contemporary catalog-adorned angler. He stood at the edge of the water casting the fly rod. The caption, attributed to the two men speaking in the foreground, read something like "He just hasn't been the same since that copy of *Sports Afield* washed ashore."

There is no doubt that the industry changes who we are as

anglers. So many of us have faith in the industry's innovation—
even if only tacitly—that we not only buy the gear we believe
we need, we place faith in that gear, counting on the technology to make us better anglers or even more responsible anglers.
Of course, to one degree or another, we are in fact dependent
on those technologies through which we fish. Our interactions
with ocean and fish would be significantly restricted without
them. Imagine trying to cast a bait to a fish or dropping a bait
to a reef or wreck without the technology of fishing line—and
imagine the details of changes to that technology since Izaak
Walton taught us how to make gut lines in 1653, and how
those changes have affected the very idea of recreational fishing.
At a minimum, the new recreational anglers' ethic requires that
we acknowledge our dependence on our technologies. More
critically, though, we must also acknowledge that as anglers
we cannot operate as isolationists. We may wish to be loners
when we fish, but the future of our fishing depends on our
willingness to work en masse. It is both naive and irresponsible
to think that regulations, policies, and politics do not affect
the ocean, the fish, and us. Each of us has the responsibility to
dedicate time to understanding those policies, to participate in
democratic processes that inform and authorize those policies,
and to adhere to them as prescribed, even when we disagree
with them. Similarly, we must acknowledge that the industry,
the capital of fishing, provides not just our technologies of
access, but also the financial support to advocate on behalf of
recreational angling within the conversations that lead to local
and federal policies. That is to say, nonparticipation is an antiethical position for recreational anglers; resistant participation,
however, is necessary when framed within the boundaries of
the democratic process.

We see the consequences of such activism in historic

moments such as the 1994 Florida net ban vote. Twenty years after the vote, Karl Wickstrom, who spearheaded the net ban campaign in the early 1990s, told Florida's *Sun Sentinel*, "Some people say it's the most important fisheries reform in history."[2] For many, Wickstrom's claim echoes a positive vision of the change brought about by the net ban; for others, such as the net fishermen who had to find new ways to earn a living, the truth in Wickstrom's statement is not read as positive, but as having an importance grounded in a significant change in their livelihoods. The importance of that vote, however, resounds now in how the historical lessons we learned about the net ban will affect future legislative and cultural responses to similar situations. That is, the real importance of the net ban will be what we ultimately learned from the moment and how that historical lesson will inform the next round of similar conversations. For example, North Carolina's 2014 proposed net ban echoes the Florida net ban and sits unresolved in a tangle of local conflict that contributes to the larger fisheries knot.

North Carolina's recreational anglers, like those in many other states, face ideological and data-driven conflict with local commercial harvesters. These conflicts should be of great importance to recreational anglers across the country given North Carolina's prominence as a recreational fishing destination. According to North Carolina Sea Grant, the state is fourth in recreational harvest and the third most popular destination for recreational anglers. Antithetical to the success of recreational angling, North Carolina's commercial fishing industry has deteriorated significantly over the last several years, falling to a ranking of fifteenth in commercial harvest nationwide.[3] At the center of the net ban sits the economic conflict over ownership of North Carolina's fish stocks, with commercial harvesters pointing to the net ban as the death knell for commercial harvest in the state. Recreational anglers, on the other hand, see the ban as contributing to a longer-term sustainability of both the inshore fishery and the economy that will continue to grow with the increase of recreational participation and tourism.

Tabled since 2013, the North Carolina net ban has been a fishing knot that legislators have seemingly not wanted to face. In May 2016, though, Representative William Richardson (D-Cumberland) filed a new bill, H.B. 1122, the Limit Marine Net Fishing Bill, which sits as of this writing with the North Carolina House Committee on Rules.[4] If passed, the bill would put to popular vote the question of whether to ban gill nets for harvesting from North Carolina's coastal waters. Interestingly, the language of H.B. 1122 turns to a rhetoric of ownership and long-term sustainability, claiming that "the marine resources of the State of North Carolina belong to all of the people of the State and should be conserved and managed for the benefit of the State, its people, and future generations. To this end, the General Assembly enacts limitations on marine net fishing in coastal fishing waters to protect saltwater finfish, shellfish, and other marine animals from unnecessary killing, overfishing, and waste."[5] In essence, the bill makes explicit that inshore netting is detrimental to the sustainability of the state's fisheries. This claim runs headlong into questions of scientific data, cultural longevity, and economic evidence, taking up some of the most problematic debates in fisheries management. The bill proposes not a legislative vote to determine the net ban policy, but a popular vote to give North Carolinians the opportunity to determine the future of their fisheries in the same way that Florida voters had a say in their state's net ban.

Ultimately, these two cases reveal deeply tangled problems in marine management, marine ethics, and fishing cultures within a US framework. However, this knot is minuscule compared with the snarl of global ocean and fisheries management. In fact, given enough prodding—and perhaps bourbon—I'd consider that issues such as net bans addressed at the state level risk failure to recognize local fisheries issues as part of larger ecological systems. For example, redfish and trout are two of the primary species addressed in the Florida and North Carolina net bans. These fish are talked about as "local" fish or as "North Carolina's fish." However, many of the fish affected by commercial harvest or net bans or other state-water-focused legislation are in fact only temporarily North Carolina's fish, or Florida's fish, or Connecticut's fish, and so on. Redfish migration, for example, might reveal redfish populations as transient, local to North Carolina only during certain periods of the year and to Virginia's or South Carolina's waters at others. Netting fish in North Carolina might deny Virginians access to redfish that would have migrated through Virginia had they not been netted in North Carolina, and vice versa. That is, issues such as net bans are national issues and, even more likely, global issues and might be better managed as such. However, to say so only tightens the knot further, confounding anglers who pick and pull at the strands to try to understand the connections. To point to net bans as a national or global issue simultaneously works against states' rights to manage local fisheries. Pulling on the strands to support states' rights results in reduced resources to examine global connections, and so on.

Problematically, though, there are no ways to offer solutions to any of these issues independent of the others to which they are bound. Nor can we even attempt to examine the smaller facets of the larger issues, as their complexity compounds with each strand examined. There is no knot within the knot that can be identified and tugged on independently of the whole, as this political, cultural, and economic ganglion is itself a self-sustaining system. Loosen one part of the knot, and other parts tighten. This is not a matter of simple connection between pieces within the system. Rather, this is a knot tied of a single line, not in a beautiful Celtic fashion but in a helter-skelter scramble that perplexes attempts to unravel it.

In the face of this perplexity, though, a new anglers' ethic remands the knot to the individual angler—not to untangle as a whole, but to treat as a dynamic system in which we must each learn to thrive. The knot is reimagined as a return from the chaos of global fisheries politics to the role of the individual recreational angler. The knot, that is, emerges as the metaphorical challenge of the saltwater fishing life. As wayfarers, we learn to navigate this knot, not bound to it as permanent form, but participating in its perpetual retying. The knot gives us pause to continually amend the saltwater fishing life and our relationship with the global ocean. The knot reveals not merely the tangles, the difficulties, and the frustrations, but the moments of stability and function within the bedlam of ocean politics. It is at once our culture, our economy, and our ecology. The knot binds, connects, and unites. The knot is our totem and our challenge, our vitality and enigma.

10

The Firebird

Belief in advancing disaster is actually part of a trust in salvation, whether deliverance is expected by sacred or profane revelations, through revolution, dramatic scientific breakthroughs, or religious rapture.

Donna Haraway, *Modest_Witness@Second _Millenium.FemaleMan©_Meets_OncoMouse™*

Admittedly, I read a lot about fishing, about the ocean, about the policies and philosophies that govern our interactions with them. I am certain that I want to write about fishing and the ocean because I read so much about them in conjunction with my time engaged with them. In my fantasies of the perfect life, I envision a life on the water, punctuated by exquisite food and drink, shared with my family and a handful of friends. My time occupied, I daydream, with countless lionhearted hours chasing fish across the global ocean. Downtime spent lost in the words of those who write about fishing, about the ocean, about adventure. I want to live in the fishing library, the archive of fishing stories. I want to live with the stories of the anglers.

Historically and comparatively, freshwater fishing, particularly fly-fishing for trout, receives substantially more written attention than does recreational saltwater fishing. Though many might point to *The Old Man and the Sea* as the quintessential fishing story, tales of trout and rivers dominate the

angling literary canon. Readers are more apt to know Nick
Adams than Santiago, to know Sparse Grey Hackle than Zane
Grey, to know *A River Runs through It* or *The River Why* than
The Book of the Tarpon or *Crunch and Des*. There is no question
for me that part of the new anglers' ethic must include critical
attention to the ways in which we communicate about fishing.
That is, to be reductive, within the historical framework of all
that has been written about fishing and the ocean, we need to
be alert to the stories we tell, the stories of the anglers. This
includes how we articulate our activisms in addressing policy
and regulation, as well as the fictions we write and the fishing
stories we recount.

With as much time as I spend reading and writing about
fishing, I am often asked what my favorite fishing book is.
It is a question I find as frustrating—if not as naive—as the
question I mentioned at the beginning of this book: "What's
the biggest fish you have ever caught?" Asking about a favorite
book is to reduce all books to a single demarcation. Books are
as different and diverse as fish, and the experience of each so
unique that asking about one's most-prized read is pointless.
There are books I love for the characters, books I love for the
instruction, for the lesson, for the contemplation, for the inspiration, for the data, for the science, for the speculation, and so
on. Some I love simply because of who wrote them. Naming
favorite books is tantamount to admitting you have a favorite
child.

Saying so, though, may be nothing more than a trite dis-
claimer, an academic attempt to free myself from the responsibility of claiming favorites. (Don't parents always really have a
favorite kid even if they won't admit it to themselves or the kid?
Mom always did love you best.) Nonetheless, I acknowledge that
if there were one book that I would recommend to all anglers,

one book that all anglers should read, one book that summarized the new recreational anglers' ethic better than most and that truly conveyed what it means to be a recreational angler, one book that I read and reread, then that book would be Dr. Seuss's *McElligot's Pool*. It is a book that ought to be considered alongside *The Compleat Angler* as an angler's bible.

McElligot's Pool is a book of optimistic deliberation and prophetic warning. It is at once a dreamer's book and a conservationist's book. Published in 1947, *McElligot's Pool* predates *The Lorax* by twenty-four years and is Dr. Seuss's preliminary foray into conservationist politics aimed at younger audiences—a savvy move to initiate cultural change early in their lives. The book tells the story of an idealistic boy, Marco—who appeared ten years earlier in Dr. Seuss's first book, *And to Think That I Saw It on Mulberry Street*—who enthusiastically hypothesizes the possibilities of fishing a small pond on Farmer McElligot's property. He imagines the pond connected to a vast system of aquifers and underground rivers—connected, he fantasizes, across the planet, perhaps even to the ocean. He dreams of an untold number of exotic specimens he might hook and catch in the pool. His aquatic menagerie is a hallmark of Seuss's creative genius and his prognostic ability to imagine alien ichthyology ahead of anyone, to ask us to consider the very idea of what we understand the essence of "fish" to be. I know I want to catch a fish that goes "GLURK," and a "Thing-A-Ma-Jigger"; "A fish so big, if you know what I mean, That he makes a whale look like a tiny sardine!"

Marco's hopefulness resonates with a young angler's idealism and enthusiasm. Marco personifies the very magic that Orwell's George Bowling ponders regarding kids enamored of fish and fishing tackle. Marco wants to catch fish. However, Marco's enthusiasm faces a challenge from Farmer McElligot, who calls Marco a fool for trying to catch fish in the pool. The pool, he tells Marco, is too small and filled with the junk others have dumped in it over the years: boots, bottles, and cans. Yet McElligot does not dissuade Marco, who recites for McElligot an inventory of all the fish he imagines might be accessible through the pool and its connections to other waters.

Many have compared *McElligot's Pool* with Seuss's *The Lorax* and it has been touted—to a lesser extent than *The Lorax*—as a critical environmentalist text, but it certainly has never earned the credence or popularity of *The Lorax*. However, it ought to be a book that every angler knows. Parents should bestow it on children in the same way we bequeath copies of Walton, heirloom knives, or antique lures. We should adopt Marco as the patron saint of the optimistic angler and tattoo his image alongside our other icons, totems, and logos. On bad days of fishing (are there such things?), we should ask ourselves, "What Would Marco Do (WWMD)"? By the same token, anglers ought to adopt September 12 (the day the book was first released in 1947) as "take a kid fishing" day to promote the recreational fishing culture in harmony with Marco's hopefulness.

McElligot's Pool, critics have argued, is a commentary about biodiversity, ocean sustainability, pollution, private lands rights, human-nature relationships, marine ecology, and so on. We might ask why *McElligot's Pool* does not enjoy the same degree of popularity as *The Lorax* as a conservationist text. We could point to the publication of *The Lorax* in 1972, just two years before Aldo Leopold's *A Sand County Almanac*, during a cultural moment in which the environmentalist movement was establishing itself as a powerful political force in the United States. Leopold's "Thinking like a Mountain" confirmed that American environmentalist thinking would be grounded in a terrestrial logic and ethos. If the United States was not quite ready to embrace environmentalism in 1972, then it was certainly not ready to take up a culture of ocean protection.

Two weeks before completing the initial draft of this book, I happened to wander into the Animazing Gallery in Las Vegas. A series of portraits of Marvel's Avengers painted in vibrant hues caught my eye, and I wandered in driven by my never-ending, though amateurish, comic geek's aesthetic. As I was pursuing the comic art, I noticed a few paintings and sketches by Charles M. Schulz, including a magnificently simple charcoal sketch of Snoopy in his Red Baron persona. I spoke with the curator of the gallery about the piece and some of the other unique pieces by Schulz and Tom Everhart, the only artist ever licensed to paint *Peanuts* characters. I suppose the curator overestimated my expertise in comic art, but he invited me to see a piece he had in a locked room a floor above. I was awed by what I saw. I will not geek out here the way I did in the gallery about the remarkable examples of art by Nicoletta Ceccoli, Fabio Napoleoni, Virgil Ross, Chuck Jones, and a slew of others; though it is difficult not to (Virgil Fucking Ross, for gods' sake!). However, even in this loggia of art, I found myself drawn (ha!) to one painting in particular. It is a picture of massive crashing waves. Its hues of blues punctuated by an orange sun setting or rising and the white details of sea foam. The painting captivating in its depiction of the offshore ocean.

At nearly the perfect optical center of the painting, a thin, rough-edged, fire-red line creates a slow, down-turned curl across the right side of the painting. The line, as one's eyes follow it across the painting, reveals itself to be the long tail feather of a stylistically unquestionable Seuss-drawn bird. A flaming red bird, flying above the cobalt churn of the ocean. A *Firebird*, as the title of the painting reveals. A bird that would be nothing more than a red bird in the painting without the marked peculiarity of the Seussian tail. A long tail, in the vein of statistical theory, suggesting the bird's own distance from its place above the ocean. The painting is one of a pair that Seuss made; the second, entitled *Freebird*, uses similar elements, though the ocean is depicted a bit more calmly in color and action and the bird is a more subtle yellow. There is an effortless caprice to the Freebird's flight, its tail more splayed as it flies to the center of the painting. *Firebird*, on the other hand, depicts steadfast movement away from the center. The Firebird determined amid the chaos, yet uncertain in its destination. Both Freebird and Firebird are wayfarers, their navigation tempered by the moment.

The gallery's copy of the painting was one of about 850 prints made and authorized by the Seuss estate (they are available through a number of dealers nationwide). I had not known the painting prior to seeing it in the gallery and learned of its partner only through my research after leaving Las Vegas. I might have anticipated it knowing that Seuss had often tried to be taken seriously as a painter and that the window in his La Jolla home studio, where he worked for more than forty years, provided a 180-degree view of the Pacific Ocean. His relationship with the ocean tempered his professional and personal life. "I can't imagine Ted really being productive without that view," his wife, Audrey, has been quoted as saying.[1] Seuss composed the geminate paintings in 1974, and the rumor is (according to Seuss's publisher) that the two hung opposite the panoramic window from 1974 to 1975. The prints were commissioned after his death in 1991.

The painting captivated—captivates—me. Without the presence of *Freebird* in the gallery, *Firebird* dominated my visual lust. I probably spent an hour in that small room in the Animazing Gallery ruminating on *Firebird*. Discovering the painting—or the print, more accurately—as I was finishing this manuscript left me thinking there in Vegas, with the

Pacific Ocean more than two hundred miles away, about living the saltwater fishing life, about adopting a new anglers' ethic. Tucked away there in that small, windowless, second-floor room amid the canvases, my mind was lost at sea—lost not only in the Firebird's sea or Marco's sea, but in that imagined sea where those of us who fish throughout our lives reside.

I thought liquidly. I tried to grasp the sublimity of the ocean in Seuss's blues and whites. I wanted to understand the Fire-bird's wanderings. I conjured up the air the bird breathed and the fish that swam below it, fish that were necessarily Marco's fish. This was the same ocean—it had to be—in which Marco envisioned transoceanic fish swimming through underground rivers and aquifers to his picayune pool. I looked to the blue waves, wanting to discern a sailfish or marlin or other familiar game fish hidden in the blue mayhem, but I knew none would appear, as this was Marco's ocean, Seuss's ocean, where no such fish might dwell. Perhaps because of their fantasy, or maybe because of their reflection of an oceanic obsession, *Firebird* and *McElligot's Pool* infuse oceanic sublimity with optimism in saltwater suspension. They tease at why we continue to fish, why we continue to turn our eyes to the sea.

Six months prior to my overdue discovery of *Firebird*, I encountered a different work of optimism—or at least one in which I strive to see optimism—that is likely to change the world of recreational saltwater fishing for US anglers. In April 2017, US representative Garret Graves (R-LA), in partnership with Representatives Gene Green (D-TX), Daniel Webster (R-FL), and Rob Wittman (R-VA), introduced H.R. 2023, the Modernizing Recreational Fisheries Management Act of 2017 (MFA). This act, which a bipartisan list of twenty-four congressional representatives ultimately sponsored, seeks to revise the Magnuson-Stevens Fishery Conservation and Management Act

(MSA) with an eye to recreational fishing. As I have pointed out earlier, the MSA has been the doctrine for fisheries management since 1976, and it has consistently set precedent that privileges commercial harvest over recreational fishing. The MSA has been amended a handful of times since it was initially passed, but it remains remarkably flawed and skewed toward commercial harvest. The MRFA seeks to change this.

Fundamentally, we can say—as I have explained throughout this book—the biggest flaw of the MSA is that it conflates recreational fisheries management and commercial fisheries management, addressing them as the same thing, not accounting for any difference at all. They are, however, very different—culturally, politically, ecologically, economically, philosophically, and ethically. H.R. 2023 seeks to change this. The act opens with what may be one of the most important statements made to date about fisheries management:

> While both provide significant cultural and economic benefits to the Nation, recreational fishing and commercial fishing are fundamentally different activities, therefore requiring management approaches adapted to the characteristics of each sector.[2]

Amen. It is about time this was said this clearly. If nothing else comes of this act, this statement alone opens the door to an entirely different approach to legislative thinking about fisheries management.

To provide even more hope, the act goes on to address a second critical flaw not only with the MSA but with the entire mentality that has driven fisheries management: a flat-out failure to use, collect, and employ reliable scientific data to manage fisheries. H.R. 2023 makes it clear from the outset that not later than 60 days after the date of enactment of this Act, the Secretary of Commerce shall enter into an arrangement with the National Academy of Sciences to conduct a study of the South Atlantic and Gulf of Mexico mixed-use fisheries—

(1) to provide guidance to the South Atlantic Fishery Management Council and Gulf of Mexico Fishery Management Council on criteria that could be used for allocating fishing privileges, including consideration of the conservation and socioeconomic benefits of the commercial, recreational, and charter components of a fishery, to a Regional Fishery Management Council established under section 302 of the Magnuson-Stevens Fishery Conservation and Management Act (16 U. S. C. 1852) in the preparation of a fishery management plan under that Act;

(2) to identify sources of information that could reasonably support the use of such criteria in allocation decisions; and

(3) to develop procedures for allocation reviews and potential adjustments in allocations based on the guidelines and requirements established by this section.[3]

Clearly, the Southeast region and Gulf Coast region have been singled out at the beginning of the act because these councils have consistently flubbed red snapper fishery management so badly that if intervention is not immediate, the problem of red snapper harvest rights will expand beyond the ability to create feasible management plans. Thus, H.R. 2023 starts to address some of the biggest problems in recreational fisheries management, and that is a great step in the right direction.

Three months prior to the introduction of H.R. 2023, though, Representative Don Young of Alaska, who had proposed H.R. 1335 in 2015 to amend the MSA, introduced H.R. 200—the Strengthening Fishing Communities and Increasing Flexibility in Fisheries Management Act. The bill would reauthorize the MSA through 2022 with some revisions:

> Revisions are made to: (1) requirements for fishery management plans for overfished fisheries; and (2) catch limit requirements, including by authorizing Regional Fishery Management Councils to consider changes in an ecosystem and the economic needs of the fishing communities when establishing the limits.
>
> To distinguish between fish that are depleted due to fishing and those that are depleted for other reasons, the term "depleted" replaces the term "overfished" throughout the MSA.
>
> Fishery impact statements must analyze the impacts of proposed actions in fishery management plans on the quality of the human environment.
>
> The National Oceanic and Atmospheric Administration (NOAA) must publish a plan for implementing the Cooperative Research and Management Program.
>
> The offshore jurisdiction of Louisiana, Mississippi, and Alabama is extended from three miles to nine miles for the recreational management of red snapper.
>
> The Gulf States Marine Fisheries Commission must conduct all fishery stock assessments used for management purposes by the Gulf of Mexico Fishery Management Council for the fisheries managed under the Council's Reef Fish Management Plan.
>
> Any commercial catch share allocation in a fishery in

the Gulf of Mexico may only be traded by sale or lease within the same commercial fishing sector.

NOAA must develop a plan to conduct stock assessments for all fish for which a fishery management plan is in effect under this bill.

Additionally, NOAA must develop guidelines that will incorporate data from private entities into fishery management plans.

Clearly, the proposed bill addresses many of the issues surrounding catch shares, including limits on who can legally own catch shares within the Gulf of Mexico.

In December 2017, Representative Young developed an "amendment in the nature of a substitute," which allows parts of the bill to be replaced. In doing so, he incorporated nearly all the language of Representative Graves's MFA into H.R. 200, essentially redefining the MSA revision in the language of the MFA. On December 13, 2017, the House Committee on Natural Resources approved H.R. 200 with those revisions.

On February 28, 2018, the US Senate weighed in on the matter when the Committee on Commerce, Science, and Transportation overwhelmingly approved Senate Bill 1520, the Modernizing Recreational Fisheries Management Act, the version Representative Young used in his revisions.

In July 2018, as I make final revisions to this manuscript for publication, the House of Representatives voted on and passed H.R. 200 with a vote of 222 yeas to 193 nays. This vote marked the first time revisions to the nation's primary marine fisheries laws would include concerns of recreational anglers. Leaders from across the recreational fishing and boating industries praised the passage of the bill. Patrick Murray, president of Coastal Conservation Association, put it best: "We are grateful to our champions from both sides of the aisle in the House for recognizing the needs of recreational anglers and advancing this important fisheries management reform. This is truly a watershed moment for anglers in our never-ending quest to ensure the health and conservation of our marine resources and anglers' access to them."[4]

The bill, identified as Senate Bill 1520, has now been placed on the Senate Legislative Calendar under General Orders, and the recreational fishing community awaits a call to vote. Online chatter about the passing of H.R. 200 and the possibility of S. 1520 carries anticipation and hope. I cannot help but read the hope in H.R. 2023, S. 1520, and H.R. 200 now without thinking of it in the shadow of *McElligot's Pool* and *Firebird*. Together, these texts convey not necessarily Marco's youthful idealism, but a kind of encouragement that recreational saltwater anglers have begun to establish a stronger sense of community and are becoming cognizant of that community's reach and voice. Unquestionably, that voice has been championed not just by these congressional leaders, but by conservation groups and industry groups working to galvanize the voices of recreational saltwater anglers.

Aldo Leopold, in his classic essay, taught us to think like a mountain, to see connections within an ecosystem and think of ourselves not as individuals but as parts of an intricate ecological web. Thinking like a mountain gave rise to the evolution of environmental politics, land ethics, and other notions of sensible approaches to land management. Many writers have since adapted Leopold's approach, asking us to think like chimpanzees, chickens, wolves, eagles, rivers, plants, savannas, and so on. However, few have asked us to think like fish or to think like oceans. To think like the unknown is difficult; it is alien.

I am not certain that thinking like fish or ocean will teach us how to interact with fish and ocean. There is more value, it seems to me, in thinking like saltwater anglers—in adapting oceanic perspectives that are willing to modify fluidly to changing conditions, in adjusting a conservationist ethic that is steeped in contemporary contexts—political, economic, ecological, cultural, and more.

Our responsibility is to acknowledge that the debated arenas regarding ocean sustainability have histories and contexts that affect people's lives and the lives of nonhuman organisms (fish, among others). It is our responsibility to understand that the changes we make in technology, philosophy, culture, and policy intrinsically retain significant connection with those histories, as well as with the yet-to-be-known future of sustainability of the world's ocean and its populations—including the human. It is our ethical imperative not just to fish or not just to be able to say "I fish," but to articulate what that means historically, culturally, and personally—to know why we do what we do and to understand our impact.

I return to the old fishing adage *piscator non solum piscatur*, which means "it is not all of fishing to fish," or more to the point, "there is more to fishing than catching fish." Traditionally, we understood this adage to refer to the contemplative nature of fishing, to the more ephemeral aspects of the angler's connection to the natural world—perhaps even to what many have referred to as the religion of fishing. We see the sentiment echoed in the words of Thoreau that I quoted earlier: "Many men go fishing all of their lives without knowing that it is not fish they are after." However, more recently, this adage seems even more applicable to recreational saltwater fishing in that fishing has come to be about much more than catching fish. It is a way of life, and like any way of life in a capitalist system, it comes with the trappings of its cultural setting. There is indeed more to fishing than catching fish. Recreational fishing is about privilege and responsibility; it is about technology and science; it is about politics and ethics; it is about life and death. It is about fish, hook, and human, certainly, but only in its most rudimentary understanding. It is, with an eye toward transparency, so much more. We find ourselves at a moment when acknowledging such is imperative to our relationship not just with fishing and fish, but with the global ocean as well.

Sylvia Earle acknowledges that "there is a tendency for most people to be complacent about the ocean and ocean life, and to regard the sea and all that it provides as somehow not as relevant to our everyday lives as terrestrial and freshwater environments."[5] We find ourselves at a moment when the future of the global ocean demands that we—as recreational anglers, as responsible stewards of the ocean, as democratic citizens—now place the future of the world's ocean and its inhabitants at the fore of our new ethical imperative. To fail to do so is to damn not only our communal pastime, but our very sustainability.

FISH ON.

Postscript

Toward a New Recreational Saltwater Anglers' Ethic

A Manifesto

Throughout this book, I have argued for a new saltwater recreational fishing ethic, as both philosophy and practice. Our cultural and practical pasts can no longer answer the needs of the global ocean. The future health of the ocean and of recreational saltwater fishing depends on our ability to articulate a new code of ethics, adhere to those ethics, and propagate those ethics to the end of maintaining an ethically credible and active community of anglers dedicated to the protection of the ocean and its inhabitants. Our code of ethics requires upgrade. What follows is an attempt to galvanize the key points of such an ethic as I have articulated them throughout this book. This is a call to change, a call to a new ethic. It is an initiation. It is at times polemic, and it may be uncomfortable. It is, after all, a call to change. But change is necessary if recreational anglers—saltwater and freshwater—are to contribute to the protection of our communal waters and our deeply treasured pastime.

This is not a beginning, but the next phase in revising the ethical practices to which we have become accustomed and which the current historical moment demands as we rethink our position in the world as anglers. I anticipate that these values will grow and mutate, and I anticipate that many of them will be critiqued and challenged. Mostly, though, I intend them as fluid, context driven, and adjustable.

No matter. What they must not be is dismissed. The future of the world's ocean and the future of recreational saltwater fishing depend on our willingness to engage and instigate evolutionary and revolutionary changes to our ethical and cultural practices.

What follows is not complete. It does not address the specifics and details of all that is involved in recreational angling. For example, it does not address the use of specific kinds of equipment in specific situations, despite such choices being directly bound up in the ethical decisions we make about fishing.

This, then, is a manifesto, a pronouncement of a new ethic for saltwater anglers.

THE 2018 RECREATIONAL SALTWATER ANGLERS' MANIFESTO

I. Introduction

1. Saltwater recreational fishing is not a right. It is a privilege. It is an activity of the privileged. Certainly, there are rights affiliated with recreational angling, but they manifest only after we have attained the privilege to fish recreationally. That privilege is primarily a result of economic privilege and the ability to afford recreation. Most of the world's population lacks this privilege.

2. To fish recreationally is to embrace our privilege. We must willingly acknowledge this to ourselves each time we fish, expressing to ourselves and to the world our gratitude for such privilege. We are fortunate to live in a country where the very idea of a $27 billion recreational fishing industry can even be fathomed, let alone be part of everyday life.

3. With such privilege comes great responsibility.

4. Given that there are eleven million saltwater anglers in

the United States (not to mention the total forty-nine million anglers that include freshwater and saltwater anglers), the recreational saltwater fishing community needs to begin to work together more closely to develop regional and national platforms and positions that lend a communal angler voice to policy making.

5. This is the moment to remake what it means to be an angler and to fish. This is the moment for American recreational anglers to embrace changes in ethical positions that will influence angler actions.

6. Only by establishing flexible and condition-malleable ethical guidelines informed by scientific data will we ever be able to maintain a culture of ethical angling that will be able to work to sustain the world's ocean while promoting the cultural value of recreational saltwater angling.

7. The moment we condemn scientific research as inconvenient or the moment we assume our individual right to harvest trumps the scientific evidence, we have surrendered our ethical position.

8. The contemporary anglers' ethic requires accounting for three primary facets of recreational fishing: (a) the actions of the angler, (b) the effect on the fish, and (c) the impact on local and global ecosystems.

II. Angler

1. All recreational anglers—saltwater and freshwater—must adhere to the strictest care and empathy in harvest methods.

2. The only fish to be harvested are those to be eaten, those to be used as bait to catch other fish, or those to be used for scientific research.

3. The fish you kill is not your fish, and it is not my fish;

it is our fish. The fish I kill is not my fish; it is not your fish. Assumed possession by the act of capture cannot be the driving philosophy behind an assumed right to kill. The recreational angler must now acknowledge a collectivity to harvest ethics. Hence, the very idea of what it means to fish recreationally should now imbue a catch-and-release mentality.

4. Recreational fishing and sportfishing must be understood as synonymous with catch and release. The right to decide to keep fish for nourishment should never be eliminated (or even questioned) as a recreational angler's right, but angler thinking should frame such decisions within a new culture that understands harvest as the minority objective, not the primary objective. There is no need to kill for any other reason; to do so is not an inherent human right. This includes, by extension, the unnecessary slaughter of recreational bycatch, whether bait or trash fish. None of us can assume we are well off enough to discard any living marine organism, because we can no longer assume an infinite wealth of oceanic resource to provide better than what we dispose, nor can we operate as individuals making claims to such wealth.

III. Advocacy and Communication

1. The ocean faces desecration from many provenances. Our awareness and involvement in the nation's ocean management requires that recreational anglers be alert to the decisions legislators make affecting all fishing, not just fishing of local concern.

2. Eleven million saltwater anglers have the potential to be a significant political voice in how fisheries policies are addressed.

3. We may wish to be loners when we fish, but the future of our fishing depends on our willingness to work en masse. Civic participation is critical. Lending voice to the eleven million is essential.

4. It is both naive and irresponsible to think that regulations, policies, and politics do not affect the ocean, the fish, and us. Each of us has the responsibility to dedicate time to understanding the policies that affect fishing, to participate in democratic processes that inform and authorize those policies, and to adhere to them as prescribed, even when we disagree with them.

5. Part of participation and awareness grows from involvement with conservation and industry organizations.

6. Anglers must acknowledge that the recreational fishing industry provides not just our technologies of access, but also the financial support to advocate on behalf of recreational angling. Support of the industry is support of recreational angling.

7. Refusal to participate in the democratic processes associated with fisheries management policies is an antiethical position for recreational anglers. Resistant participation, however, is necessary when framed within the boundaries of the democratic process.

8. Recreational anglers—and those that represent them in industry, government, and culture—must be critically conscious of the ways in which we communicate about fishing. Within the historical framework of all that has been written about fishing and the ocean, we must be alert to the stories we tell, and the culture we communicate. This includes how we articulate our activisms in addressing policy and regulation, as well as the fictions we write and the fishing stories we recount.

9. For-hire captains are obligated to teach. They are our

emissaries, our professional class. Guides contribute to the conservation of ocean and fisheries not only by their actions but through their teaching. Guides are obligated not only to put their clients on fish, but to use those experiences to teach clients about the ethics of recreational angling and the conservation efforts bound to the activity.

10. Despite some fundamental differences, recreational anglers and commercial harvesters must collaborate to find solutions to the increasing limits and restrictions imposed on harvest approaches, most notably catch-share strategies. Together, these groups are more likely to develop feasible long-term conservation and harvest strategies than they are if they continue working in opposition.

IV. Fish

1. If recreational anglers are to move toward practices that promote ocean sustainability, then we must first move toward ways of thinking about fish and fisheries beyond mere economic frames. This includes, primarily, understanding the role of every organism in local and global ecosystems.

2. All fish are a part of the commons. Certainly, those who do not participate in recreational angling or commercial harvest have some degree of claim over fish as resources of the commons. If any fish I catch is not my fish, then it is, in fact, as much the nonangler's fish as it is the angler's. This, however, does not authorize the nonparticipant to determine the harvest ethic of the participant. It only authorizes the nonparticipant to become a participant. Likewise, we must consider whether the nonparticipant has the ethical right to assign a proxy

participant or whether others can claim a role as a proxy participant without nonparticipant ratification, as is the case with commercial and industrial harvest.

3. Anglers must adjust the very underpinnings that have allowed us to imagine ocean and fish as property to be claimed, distributed, and owned. Instead, the anglers' ethic ought to recast the ocean as a global region critical to all humans and stewarded by all humans.

4. There is no such thing as trash fish.

V. Ecology

1. All discussion of anglers' ethics—no matter how criticized—must account for individual angling actions within the scope of global ocean conservation. That is, we must acknowledge that any harvest we engage has a larger, global effect. Individually, this may be difficult to grasp, as perceptually it is likely that any individual angler's harvest may be assumed insignificant in relation to global fisheries or ocean conservation. However, given the vast number of recreational anglers now fishing in the United States, we cannot dismiss the individual angler's impact as being that of only the individual angler; it is that of the individual angler compiled with that of every other individual angler. The individual angler must be accounted for as a member of a collective. Thus, the individual angler's minimal harvest is compounded across the recreational angler population and can have a noticeable effect not just on fishery populations, but on every aspect of human engagement with the ocean. We may fish alone, even seek the solitude of the angler's devotion, but our actions aggregate.

2. Catch and release is not just for individual fish; it is for larger ecosystems.

VI. Technology

1. We now need to recognize that fishing itself is a technological intervention, that without some form of innovation—be it Paleolithic fish traps or thermoplastic elastomer lures—the very act of fishing would not be possible.

2. To be a recreational angler requires that we be beholden to our technological developments to some degree. That degree, however, may be personal, or it may be institutional. Making such choices is central to the recreational anglers' ethic.

3. The privilege of recreation is provided by a host of technologies that make access available, and it is also a privilege to be able to afford those technologies and the time in which to use them. Just as we talk about the digital divide and the effect access has on education and commerce, so too must we acknowledge the effect the technological divide has on angling access.

3. All actions tied to fishing—use of boat fuel, food wrappings, fishing lines and lures, litter, pollution, boat emissions, and so forth—contribute to the footprint of the angler.

4. Our fishing technologies have global ecological impact beyond their use to extract fish and their disposal as waste. For example, the mining of the minerals needed to build navigational electronics or the petroleum used to make plastics has global ecological impact beyond the immediate application in recreational angling.

5. Ultimately, we need to come to terms with the difficult position that in order to protect recreational saltwater angling, the fish populations and ocean environments on which recreational angling depend require more technological intervention, not less.

6. Fish populations are necessary in the global protein economy. Recreational anglers must support technological development of methods such as mariculture as an alternative to wild harvest while simultaneously finding a balance for the management of wild harvest that accounts for recreational needs and a reduced and monitored commercial harvest. All such ethics will inevitably be bound up in complex management policies, as well as deep-seated philosophies of what it means to be a recreational saltwater angler.

Notes

CHAPTER 1

1. Mitchell, *Sea Sick*, 22.

2. Brayton, *Shakespeare's Ocean*.

3. Parry, *Discovery of the Sea*.

4. Some NOAA and NMFS materials from 2014 identify the figure as twelve million. For instance, see http://www.nmfs.noaa.gov/sfa/management/recreational/recreational_fishing_initiative.html.

5. Cooke and Cowx, "Role of Recreational Fishing," 857–59.

6. Many of you may think that this number seems low, as did I when I first saw it. However, it is accurate. What provides the impression that the NRA has a larger support base is that in addition to the five million registered members, more than fourteen million people self-identify as supporting the NRA and its mission but do not have membership.

7. Southwick Associates. "Trade Insights and Market Report." 2018.

8. Keep in mind, too, that the NRA galvanizes around an identifiable piece of equipment: the firearm. The activities it supports (such as hunting) orbit the organization's mission to support "a wide range of firearm-related public interest activities" and to "defend and foster the Second Amendment rights of all law-abiding Americans." Anglers have no such object around which to unite. Rod rights? Hook rights? These objects do not carry the same investment or legal and cultural histories as firearms. Thus, finding common ground on which anglers can forge a common objective is quite different than it is for the NRA, despite the frequent reduction of hunters' rights and anglers' rights under the same "outdoorsmen" philosophies.

9. A 2015 report from the American Sportfishing Association titled *The Economic Gains from Reallocating Specific Saltwater Fisheries* indi-

cates that these kinds of expenditure differences may be measurable within regional locations, so assumptions can be made that, for example, a striper angler's expenditures are similar to those of other striper anglers, but not similar to those of a halibut angler.

10. Interestingly, given the successful introduction of chinook salmon to New Zealand, researchers tried to introduce chinook to Hawaiian waters, as well, but failed.

11. Nature Conservancy, *Fishing for Healthy Reefs: Spearfishers Target Alien Species to Help Maui Reefs*, 1998, http://www.nature.org/wherewework/northamerica/states/Hawai'i/projectprofiles/art28916.html (site discontinued).

12. Similar "roundups" are now held in Florida to remove lionfish—also an invasive—from Florida's reefs.

13. Historically, there are few moments like this in US recreational fishing. Notably, H. Bruce Franklin's book *The Most Important Fish in the Sea* describes an event in 1877 in New Jersey in which recreational anglers threatened to obtain cannons to fire on industrial menhaden harvesting boats unless the government intervened and curbed the menhaden slaughter. Similarly, in 1922 anglers in Cape May and Wildwood, New Jersey, demanded that the government run off the menhaden fleets that were destroying local fisheries or they, too, would fire cannons at the fleet. Franklin, *Most Important Fish*, 99.

14. "Gulf of Mexico Recreational Red Snapper Management," NOAA Fisheries Southeast Regional Office, http://sero.nmfs.noaa.gov/sustainable_fisheries/gulf_fisheries/red_snapper/index.html.

15. "Red Snapper Proposal," Gulf States Red Snapper Management Coalition, March 13, 2015, https://vdocuments.mx/red-snapper-proposal.html.

16. Mike Leonard, email message to author, October 9, 2017. Cited with permission.

17. "Red Snapper Scheme Could Destroy Fishery," April 1, 2015, Seafood Harvesters of America, https://www.seafoodharvesters.org/2015/04/01/red-snapper-scheme-could-destroy-fishery/.

18. "Congressional Sportsmen's Caucus Members, State Agencies Discuss State-Based Solution to Red Snapper," Congressional Sportsmen's Foundation, April 30, 2015, http://congressionalsportsmen.org/the-media-room/news/congressional-sportsmens-caucus-members-state-agencies-discuss-state-based-

19. "H.R. 1335: Strengthening Fishing Communities and Increasing Flexibility in Fisheries Management Act," Representative Don Young, March 4, 2015, https://www.congress.gov/bill/114th-congress/house-bill/1335/text.

20. "Legislation Advances to Benefit Saltwater Recreational Fishing," American Sportfishing Association, May 1, 2015, https://asafishing.org/legislation-advances-to-benefit-saltwater-recreational-fishing/.

21. Gerhard Peters and John T. Woolley, "Barack Obama: Statement of Administration Policy: H.R. 1335 - Strengthening Fishing Communities and Increasing Flexibility in Fisheries Management Act," The American Presidency Project, May 19, 2015, http://www.presidency.ucsb.edu/ws/index.php?pid=11013820.

22. Bradley Byrne, "Providing for Consideration of the Bill (H.R. 1335) to Amend the Magnuson-Stevens Fishery Conservation and Management Act to Provide Flexibility for Fishery Managers and Stability for Fishermen, and for Other Purposes," May 19, 2015, https://www.govtrack.us/congress/bills/114/hres274.

23. "H.R. 1335: The Wrong Way Forward," PEW Charitable Trust, May 22, 2015, http://www.pewtrusts.org/en/research-and-analysis/issue-briefs/2015/05/hr-1335-the-wrong-way-forward.

24. "Committee Information," US Senate Committee on Commerce, Science, and Transportation, http://www.commerce.senate.gov/public/index.cfm/about.

25. "H.R.3310 - Preserving Public Access to Public Waters Act," Rep. Ileana Ros-Lehtinen, July 29, 2015, https://www.congress.gov/bill/114th-congress/house-bill/3310/text.

26. For sailors and watermen of all cuts, the term "head" refers

to the toilet. My title is thus intended to be playful, suggesting both thoughtful considerations and the possibility of me just talking shit.

27. I should note that on April 20, 2016, representatives from both the recreational and commercial fishing industries, as well members of Congress, gathered on Capitol Hill in recognition of the fortieth anniversary of the Magnuson-Stevens Act, which President Gerald Ford signed on April 13, 1976.

28. Roberts, *Unnatural History of the Sea*, xv.

CHAPTER 2

1. In 2001 the American Fisheries Society (AFS) "officially" renamed the jewfish (*Epinephelus itajara*) the Atlantic goliath grouper, claiming that the original moniker could be understood as culturally insensitive. At the time, I spoke with a representative of the AFS about the change. It turned out that in more than fifty years, the AFS had record of only one complaint regarding the name of the fish. Likewise, when asked about the etymology of the original name, the AFS was not clear on the name's history. Through my own research, I discovered that the name has three primary origins: the first refers to the fish's flesh being considered of such high quality that it was thought to reflect Jewish customs of kosher food. Second, when Europeans began settling in the New World, the fish was so abundant in Atlantic, Caribbean, and Gulf waters that it was named as reflective of verse 1:22 in Genesis, which says, "Be fruitful and increase in number and fill the water in the seas." Third, because of the fish's size, it was thought that perhaps the fish was the fish—not a whale (translations vary)—that would have swallowed Jonah, a prophet of the northern kingdom of Israel. With this information in hand, I have resisted allowing an organization such as the AFS to dictate language use and have continued to use the name by which I grew up knowing these majestic fish, despite the sport-fishing industry's willingness to comply with the mandated change. Besides, changing the name from jewfish to goliath grouper seems even

more problematic given that the Philistine Goliath was an enemy of Israel. So, in essence, the name change might be read as a move from honoring Jewish tradition to insulting it by invoking the name of one of the Jews' greatest enemies.

2. Ballantyne, *Ocean and Its Wonders*, 1.

3. Earle, *World Is Blue*, 53–54.

4. Trujillo and Thurman, *Essentials of Oceanography*, xxx.

5. Imagine what would happen should we acquire the knowledge that would allow such realizations and claims about outer space, as well. Until that time, though, we can no longer think of oceanic space and outer space as equivalent in limit or as limitless.

6. Carson, *Sea around Us*, 14.

7. Ibid., 15.

8. Diaz and Rosenberg, "Spreading Dead Zones," 926–29.

9. Moffitt et al., "Response of Seafloor Ecosystems," 4684–89.

10. According to NASA, more than half of its infrastructure— including more than $32 billion worth of laboratories, launchpads, air-fields, test facilities, and data centers—stands within sixteen feet of sea level and is at risk of destruction with only minimal sea-level rise (see http://www.cnn.com/2015/09/04/us/nasa-launch-sites-rising-sea-levels-feat/index.html).

11. Yes, IRL can also refer to "In Real Life" in internet slang, which is ripe for play here, but that's for another time.

12. Nadia Drake, "Florida Fights to Save a Troubled Lagoon and its Once-Flourishing Marine Life," August 29, 2013, http://www.wired.com/2013/08/indian-river-lagoon-toilet/.

13. Carl Hiaasen, "Scott Clueless in Lake O Crisis," March 4, 2016, http://www.miamiherald.com/opinion/opn-columns-blogs/carl-hiaasen/article64076762.html.

14. Alan Farago, "On Florida's Massive Fish Kills, Voters Are Responsible," March 23, 2016, http://www.huffingtonpost.com/alan-farago/on-floridas-massive-fish_b_9530084.html.

CHAPTER 3

1. Paul Bagne, "Interview: Christopher McKay," July 1992, http://www.isfdb.org/cgi-bin/pl.cgi?59849.

2. Berra, *Freshwater Fish Distribution*, xx.

3. Nelson, *Fishes of the World*, 1–2.

4. Helmreich, *Alien Ocean*, 8.

5. Interestingly, the colloquial "I know it when I see it" refers to US Supreme Court justice Potter Stewart's famous consideration of obscenity in the 1964 court case *Jacobellis v. Ohio*, in which Stewart wrote, "I shall not today attempt further to define the kinds of material I understand to be embraced within that shorthand description, and perhaps I could never succeed in intelligibly doing so. But *I know it when I see it* [italics added], and the motion picture involved in this case is not that." Perhaps, then, we gain a better chance of identifying what is not a fish than we do trying to define what is.

6. This concept is perhaps best articulated in Jamie Uys's 1980 film *The Gods Must Be Crazy*, a story of an African Bushman who finds a glass Coke bottle and has no cultural reference point to understand what the alien object is.

7. In Florida, capybaras are considered an invasive species and are monitored as nonnative wildlife, thus ascribing a different legal and cultural value to them.

8. Think, too, how problematic this term has been historically when used to describe particular segments of the American population.

9. Franklin uses this phrase as the title for his 2007 book as well as for the 2001 *Discover* magazine article he coauthored with Tom Tavee.

10. Franklin is only one of many English professors whose work has been influential in writing about fish and fishing. Norman Maclean, author of *A River Runs through It*, is perhaps the most well known.

11. Read Franklin's book. Seriously.

12. Franklin, *Most Important Fish in the Sea*, 5.

13. Tavee and Franklin, "Most Important Fish in the Sea."

14. Ibid.

15. As noted in the previous chapter, in 2016, massive algal blooms in Florida resulted from overflow water released from Lake Okeechobee into surrounding estuarine waters. The algal growth was attributed to increased chemical runoff and discharge from sugar farming in areas surround the lake. These two events produced more harmful algae than any natural population of menhaden or other filter feeders could have accounted for. The blooms killed millions of fish and other animals along both of Florida's southern coasts.

16. Fairbrother, "Omega Protein Makes Good."

17. Mood, *Worse Things Happen*.

18. Cooke and Cowx, "Role of Recreational Fishing," 857.

19. Cooke and Cowx's study is deeply problematic both in the assumptions it makes regarding data use and in its inclusion of freshwater and inshore recreational harvest numbers—areas not often fished by commercial operations—within their overall recreational harvest numbers. Because commercial harvest numbers either do not account for freshwater harvest or do not distinguish between the two, the comparisons that Cooke and Cowx make are often incongruent.

20. I should note, too, that forms of local fisheries thinking are embedded in scientific research, as well. Given the remarkable complexity of fisheries issues, many studies rely on methodologies that might be understood as "scenario analysis," procedures that encourage the examination of specific situations in order to extrapolate more global speculations. For example, see the United Nations Environment Programme, *Global Environmental Outlook* 3 (London: Earthscan, 2002), and its use of four scenarios for considering unfolding fisheries management issues.

21. "HB 1577 Management of Menhaden," Virginia's Legislative Information System, 2017, https://lis.virginia.gov/cgi-bin/legp604. exe?171+sum+HB1577.

CHAPTER 4

1. To be clear, given Fanta's ties to Nazi Germany, my family always favored Nehi, as did "Radar" O'Reilly, played by Gary Burghoff, who

holds a number of patents for fishing rods and other tackle development.

2. Buie, who is credited with writing more than 340 songs in the BMI catalog, did most of his writing in his fishing trailer along a creek in Eufaula, Alabama. Fishing has long been muse.

3. In late 2015, after ninety-three years, Uncle Josh announced it would stop producing pork rind baits. The company explained that changes in the pork industry had forced the decision. Now, pigs are brought to slaughter at six months old, whereas in the past they were brought to slaughter at two to three years old. The skin and fatback of the younger pigs do not have the durability of those of the older animals; the skin is just too thin to retain the strength needed to make the baits. Though only tangentially relevant to this narrative, the Uncle Josh case does present some interesting departure points for thinking about the ecology of the recreational fishing industry.

4. "IUMA: Balaz, Joe," https://archive.org/details/iuma-balaz_joe.

5. Of course, critics of recreational fishing might contend that such a position circumvents questions about whether the very act of hooking and fighting a fish is itself ethical. I will address these concerns later, though only briefly, as I see them as irrelevant for the individual who has already entered a recreational angling frame of mind. If I were to argue that we might consider *not* fishing, I might perhaps turn to such considerations. However, not fishing is, for me and millions of others, an impossible (if not irrelevant) position.

6. The question of using live bait is problematic: on the one hand, we can see the function of bait as a strategic move to ensnare larger fish, but on the other hand, bait and bycatch ultimately serve the exact same ecological function as fish that are killed as extraneous to the recreational fishing process.

7. See, for instance, Worm et al., "Impacts of Biodiversity Loss."

8. *An Inconvenient Truth* won Academy Awards for Best Documentary Feature and for Best Original Song in 2006.

9. "You'll Never Go Fishing after Reading This," PETA, 2015.

http://www.peta.org/living/entertainment/youll-never-go-fishing-after-reading-this/.

10. Braithwaite, *Do Fish Feel Pain?*, 153–54.

11. It certainly did not hurt the popularization of the sport that alongside Grey, membership in the Tuna Club of Avalon included Bing Crosby, Charlie Chaplin, Stan Laurel, Hal Roach, Cecil B. DeMille, Theodore Roosevelt, Herbert Hoover, George S. Patton, and Winston Churchill—all what contemporary marketing folks might refer to as "influence marketing."

12. Trullinger, "New Big Fish Club," included in Valenti, *Understanding* The Old Man and the Sea, 158–59.

13. Evans, *With Respect for Nature*.

14. Arlinghaus and Schwab, "Five Ethical Challenges," 219–34.

15. Ibid., 221.

16. Ibid.

17. Ibid., 225.

18. Ibid., 226.

CHAPTER 5

1. Note that the term "catch shares" suggests the commodification of a share or portion, not equal sharing or distribution.

2. "2016 Gulf of Mexico Red Snapper Recreational Season Length Estimates NOAA Fisheries, Southeast Regional Office," NMFS NOAA, May 2, 2016, http://sero.nmfs.noaa.gov/sustainable_fisheries/lapp_dm/documents/pdfs/2016/sero-lapp-2016-04_gom_red_snapper_rec_season_2016_20160502_sero_final.pdf, pp. 3–4.

3. van der Voo, *Fish Market*.

4. *Ibid., 91.*

5. Ibid., 105.

6. Ibid., 108.

7. Ibid., 109.

8. "About Us," Property and Environment Research Center, https://www.perc.org/about-us.

9. van der Voo, *Fish Market*, 110.

10. Ibid., 111.

11. Ibid., 112.

12. Ibid.

13. Jeff Angers, "Gulf Scandal Looms for South Atlantic," February 23, 2017, http://www.sportfishingmag.com/recreational-fishing-scandal-south-atlantic#page-2.

14. "EFP Application for Snapper-Grouper Catch Share Pilot Program Pulled after Widespread Opposition," Saving Seafood, March 10, 2017, https://www.savingseafood.org/news/management-regulation/efp-application-snapper-grouper-catch-share-pilot-program-pulled-widespread-opposition/.

15. van der Voo, *Fish Market*, 112.

CHAPTER 6

1. Heru Purwanto, "Indonesia to Seek UN Support to Curb Illegal Fishing," May 6, 2017, http://www.antaranews.com/en/news/110804/indonesia-to-seek-un-support-to-curb-illegal-fishing.

2. Wolf, "Environmental Ethics," 19–32.

3. Diamond, *Collapse*, 479.

4. "Global Aquaculture Production," Food and Agriculture Organization of the United Nations, http://www.fao.org/fishery/statistics/global-aquaculture-production/en.

5. Upton and Buck, "Open Ocean Aquaculture."

6. Ibid.

7. "H.R. 4363 - National Sustainable Offshore Aquaculture Act of 2009, Rep. Lois Capps, January 4, 2010, https://www.congress.gov/bill/111th-congress/house-bill/4363.

8. "Research in Aquaculture Opportunity and Responsibility Act of 2010," Senator David Vitter, May 25, 2010, https://www.congress.gov/bill/111th-congress/senate-bill/3417.

9. Mann, "Blue Water Revolution."

CHAPTER 7

1. Ingold, *Lines*, 81.

2. Camp et al., "Angler Travel Distances."

3. Ingold, *Lines*, 101.

4. Ibid., 101–2.

5. Ibid., 75.

6. Ibid., 76.

7. Ibid., 102.

8. Swan, *In Defense of Hunting*, 80.

CHAPTER 8

1. Kurlansky, *Last Fish Tale*, 127.

2. Kurlansky, "A World without Fish," June 2011, http://www.wondersandmarvels.com/2015/10/a-world-without-fish.html.

3. "James Lovelock: 'Enjoy Life while You Can: In 20 Years Global Warming Will Hit the Fan,'" Decca Aitkenhead, March 1, 2008, https://www.theguardian.com/theguardian/2008/mar/01/.scienceofclimatechange.climatechange.

4. More images, including pictures of the plane, are available at http://www.bernieschultzfishing.com/lur_super.htm and http://www.joeyates.com/superstrike.htm or https://luresnreels.com/flart.html.

5. "H.R. 2083 - Endangered Salmon and Fisheries Predation Prevention Act." Rep. Jaime Herrera Beutler, June 27, 2018, https://www.congress.gov/bill/115th-congress/house-bill/2083/text.

6. Ibid.

CHAPTER 9

1. Garnett, *Turbines*, 172.

2. Steve Waters, "Net Ban Rescued Fishing in Florida," July 1, 2015, http://www.sun-sentinel.com/sports/outdoors/fl-waters-outdoors-net-ban-0702-20150701-column.html.

3. "Commercial Fishing," North Carolina Environmental Quality, http://portal.ncdenr.org/web/mf/commercial-fishing.

4. "Limit Marine Net Fishing," North Carolina General Assembly, May 12, 2016, https://www.ncleg.net/gascripts/BillLookUp/BillLookUp.pl?Session=2015&BillID=h1122.

5. "Limit Marine Net Fishing," General Assembly of North Carolina, May 11, 2016, https://www.ncleg.net/Sessions/2015/Bills/House/PDF/H1122v1.pdf.

CHAPTER 10

1. "An Ocean of Inspiration," The Art of Dr. Seuss, http://www.drseussart.com/firebird/#ocean.

2. "H.R. 2023: A Bill to Modernize Recreational Fisheries Management," 115th Congress, 1st Session, April 6, 2017, https://www.congress.gov/115/bills/hr2023/BILLS-115hr2023ih.pdf.

3. Ibid.

4. American Sportfishing Association. "U.S. House of Representatives Passes Magnuson-Stevens Reauthorization Bill." July 11, 2018, https://asafishing.org/u-s-house-of-representatives-passes-magnuson-stevens-reauthorization-bill.

5. Earle, *Sea Change*, xviii.

Works Cited

Arlinghaus, Robert, and Alexander Schwab. "Five Ethical Challenges
    to Recreational Fishing: What They Are and What They Mean."
    American Fisheries Symposium 75 (2011): 219–34.

Ballantyne, R. M. *The Ocean and Its Wonders.* 1874. Reprint, Cre-
    ateSpace Independent Publishing Platform, January 14, 2015.

Berra, Tim M. *Freshwater Fish Distribution.* Chicago: University of
    Chicago Press, 2007.

Braithwaite, Victoria. *Do Fish Feel Pain?* Oxford: Oxford University
    Press, 2010.

Brayton, Dan. *Shakespeare's Ocean: An Ecocritical Exploration.* Charlot-
    tesville: University of Virginia Press, 2012.

Camp, Edward V., Robert N. M. Ahrens, Chelsey Crandall, and
    Kai Lorenzen. "Angler Travel Distances: Implications for Spatial
    Approaches to Marine Recreational Fisheries Governance." *Marine
    Policy* 87 (2018): 263–74.

Carson, Rachel L. *The Sea around Us.* New York: Oxford University
    Press, 1950. Reprint, 1989.

Cooke, Steven J., and Ian G. Cowx. "The Role of Recreational Fishing
    in Global Fish Crises." *BioScience* 54, no. 9 (2004): 857–59.

Diamond, Jared. *Collapse: How Societies Choose to Fail or Succeed.* New
    York: Viking, 2005.

Diaz, Robert J., and Rutger Rosenberg. "Spreading Dead Zones and
    Consequences for Marine Ecosystems." *Science* 321, no. 5891
    (August 15, 2008): 926–29.

Earle, Sylvia. *The World Is Blue: How Our Fate and the Oceans Are One.*
    Washington, DC: National Geographic, 2009.

Evans, Claude J. *With Respect for Nature: Living as Part of the Natural
    World.* Albany, NY: State University of New York Press, 2012.

Fairbrother, Alison. "Omega Protein Makes Good on Threat to Cut
    Jobs; But It Doesn't Have To." *Bay Journal,* March 31, 2013. http://
    www.bayjournal.com/article/omega_protein_makes_good_on_
    threat_to_cut_jobs_but_it_doesnt_have_to.

Franklin, H. Bruce. *The Most Important Fish in the Sea*. Washington, DC: Island Press, 2007.

Garnett, W. H. Stuart. *Turbines*. London: George Bell and Sons, 1907.

Greenberg, Paul. *Four Fish: The Future of the Last Wild Food*. New York: Penguin, 2011.

Helmreich, Stefan. *Alien Ocean: Anthropological Voyages in Microbial Seas*. Oakland: University of California Press, 2009.

Ingold, Tim. *Lines: A Brief History*. New York: Routledge, 2007.

Kurlansky, Mark. *Cod: A Biography of the Fish That Changed the World*. New York: Penguin, 1998.

———. *The Last Fish Tale: The Fate of the Atlantic and Survival in Gloucester, America's Oldest Fishing Port and Most Original Town*. New York: Riverhead Books, 2009.

Kurlansky, Mark, and Frank Stockton. *World without Fish*. New York: Workman, 2014.

Mann, Charles C. "The Blue Water Revolution." *Wired*, May 1, 2004.

Mitchell, Alanna. *Sea Sick: The Global Ocean in Crisis*. Toronto: McClelland and Stewart, 2009.

Moffitt, Sarah E., Tessa M. Hill, Peter D. Roopnarine, and James P. Kennett. "Response of Seafloor Ecosystems to Abrupt Global Climate Change." *Proceedings of the National Academy of Sciences* 112, no. 15 (April 14, 2015): 4684–89.

Mood, Alison. *Worse Things Happen at Sea: The Welfare of Wild-Caught Fish*. Fishcount.org.uk. 2010. http://www.fishcount.org.uk/published/standard/fishcountfullrptSR.pdf.

Nelson, Joseph S. *Fishes of the World*. 4th ed. Hoboken, NJ: John Wiley and Sons, 2006.

Parry, J. H. *The Discovery of the Sea*. Berkeley: University of California Press, 1974.

Randall, J. E. "Introductions of Marine Fishes to the Hawaiian Islands." *Bulletin of Marine Science* 41 (1987): 490–502.

Roberts, Callum. *The Unnatural History of the Sea*. Washington, DC: Island Press, 2008.

Starosielski, Nicole. *The Undersea Network*. Durham, NC: Duke University Press, 2015.

Swan, James A. *In Defense of Hunting*. New York: Harper Collins, 1995.

Tavee, Tom, and H. Bruce Franklin. "The Most Important Fish in the Sea." *Discover*, September 1, 2001.

Trujillo, Alan P., and Harold V. Thurman. *Essentials of Oceanography*. 11th ed. Hoboken, NJ: Pearson, 2013.

Trullinger, R. "New Big Fish Club Is Organized, but It's Awfully Hard to Crash." *New York City World Telegram*, November 23, 1936; included in P. D. Valenti, ed. *Understanding* The Old Man and the Sea. Westport, CT: Greenwood Press, 2002.

Upton, Harold F., and Eugene E. Buck. "Open Ocean Aquaculture." Congressional Research Service. RL32694. August 9, 2010.

van der Voo, Lee. *The Fish Market: Inside the Big Money Battle for the Ocean and Your Dinner Plate*. New York: St. Martin's Press, 2016.

Wolf, Clark. "Environmental Ethics and Marine Ecosystems: From a 'Land Ethic' to a 'Sea Ethic.'" In *Values at Sea: Ethics for the Marine Environment*, edited by Dorinda G. Dallmeyer, 19–32. Athens: University of Georgia Press, 2003.

Worm, Boris, Edward B. Barbier, Nicola Beaumont, J. Emmett Duffy, Carl Folke, Benjamin S. Halpern, Jeremy B. C. Jackson, et al. "Impacts of Biodiversity Loss on Ocean Ecosystem Services." *Science* 314 (November 3, 2006): 787–90.

Index

The Wardlaw Book designation honors Frank H. Wardlaw, the first director of Texas A&M University Press, and perpetuates the association of his name with a select group of titles appropriate to his reputation as a man of letters, distinguished publisher, and founder of three university presses.